JUST-IN-TIME
IN AMERICAN MANUFACTURING

JUST-IN-TIME
IN AMERICAN MANUFACTURING

by

William L. Duncan, CPIM, CPM

Richard Perich
Publications Administrator

Published by

the Society of Manufacturing Engineers
Publications Development Department
One SME Drive
P.O. Box 930
Dearborn, Michigan 48121

JUST-IN-TIME
IN AMERICAN MANUFACTURING

Society of Manufacturing Engineers
Dearborn, Michigan 48121

First Edition

Third Printing

Library of Congress Catalog Number: 88-63971
International Standard Book Number: 0-87263-352-7
Manufactured in the United States of America

Dedication

In every organization there are a few people who constantly strive for excellence. These are not the ones who file in at seven and out at four each day, and whose only purpose in attending work is to get a paycheck. These excellent few are largely self-motivated and self-directed. They are continually trying new things, taking new approaches and applying common sense in uncommon ways (my own definition of innovation). They work incessantly at improving themselves, taking classes at night, reading, getting together with associates on their own time to enhance themselves and/or their professional performance. They are often frustrated people, because the ultimate challenge for them is not the work itself but the recalcitrance of their management peers. These people are and always have been the lifeblood of American industry.

This book is dedicated to those people. Among the ones I have encountered in my own career are:

Mark Bingaman	McDonnell Douglas Aircraft
Daniel R. Bradley	Arthur Andersen
Stanley A. Breen	Signode Corporation
Kirby Davis	Chicago Cutlery
John O. Gelderman	Kar Products
Tim Gertz	Chicago Cutlery
James E. Harl	John Deere
Craig Johnson	Ernst and Whinney
Charles Kopolous	Coopers and Lybrand
Keith LeCompte	Danly Machine
Michael Lema	Sentry Schlumberger
Scott Milligan	Wisconsin Metal Products
Jerry Shields	AT&T
Walter Siok	Danly Machine
Enoch Stiff	Trak International
Ken Rydzewski	A.O. Smith
John VanDeven	McDonnell Douglas Aircraft

TABLE OF CONTENTS

Part I

Background and Definitions

Part I

Background and Definitions

Definitions of Just-In-Time are as varied as the companies who claim to use it. Many perceive it to be an inventory reduction program, an employee involvement program, or a way to squeeze more out of external suppliers. This misconception has resulted in JIT receiving unfair criticism as misapplications and ''Black Box'' approaches generate failure and frustration among many novices.

As this text will relate, other companies—American companies—have made Just-In-Time an integral part of their business. In most cases, these companies did not approach JIT as the ''next step'' in their development, nor experiment with it before deciding on its appropriateness. In some cases, JIT was adopted as a last hope for survival. As a result, the level of committment to these concepts was high. (Harley Davidson, mentioned later, is one example.)

Characteristics of the JIT ''winners'' are usually found to include the following:

1. A correct definition; Just-In-Time must be understood as a philosophy. It may be implemented in a project or program form, but as a philosophy it must be understood as an unending change in the way business is conducted. This concept will be dealt with in more detail in Chapter 2.
2. A clear recognition and working knowledge of the elements of Just-In-Time; this usually requires a model of the philosophy, with a component relationship that supports an understanding of their sequence and structure. One such model is provided in Chapter 3. The details of JIT elements are more thoroughly described in Chapters 4 through 10.
3. Willingness to change; countless organizations have given lip service to Just-In-Time, and prematurely launched pilot projects with predictably dismal results. In most cases, when push came to shove, reluctant middle managements, who had done business in the same ways for many years, proved unwilling to give up formulas which had formerly proved success-

ful. To quote Robert D. Gilbreath in his extraordinary book, *Forward Thinking*, "Because we are living in a vortex of change, we and our companies must synchronize with it..."

4. Management committment must be at every level. As is common in improvement efforts, it is relatively easy to get dramatic results rather quickly. However, as time goes on, companies must continue their improvement efforts. If they do not improve, they will decline. It is impossible to remain static in a changing world.

Part I of this text (Chapters 1 through 10) will cover the background and need for Just-In-Time in American Manufacturing, as well as a description of its components.

THE CONDITION OF AMERICAN MANUFACTURING 1

To say the pressures experienced by American manufacturers in recent years has become enormous is to state the obvious. How to make American manufacturing competitive again in the world market has become a very popular topic in the media recently. For example, the cover story of *Business Week's* April 20, 1987 issue is "Can America Compete?" The story's subtitle reads: "Its options are a surge in productivity or a lasting decline."

In a very general sense, this statement is certainly true. When we examine output vs. wage levels over the last fifteen years in both the United States and Japan, we see clear evidence of this situation. (See *Figure 1-1*). Since about 1976, United States industrial wage levels have continued a fairly steady climb, as have those in Japan. However, industrial output in the United States virtually leveled off during this same period, while Japanese output levels increased beyond wage improvements.

General productivity, however, is really only a part of the problem. Along with not producing enough product per investment dollar, we are producing the wrong things, and the wrong quantities, at the wrong times. Manufacturing inventories have more than tripled in the United States since 1965. The sophisticated production and inventory control techniques associated with MRP II, while correct in theory, have in many cases automated the generation of the wrong requirements. Automated requirements generation systems such as Material Requirements Planning (MRP) are completely reliant on accurate bills of material, routings, lead times, inventory data, and end-item requirements. Because American manufacturing built so many intrinsic delays and queues into its operations in order to overcome inaccuracies in these areas (see *Figure 1-2*), enormous inefficiencies have become common place. MRP usually does not work well as an execution tool. In order for MRP to generate accurate shop production schedules for example, "frozen" production horizons are required which are simply not realistic in the dynamic world of end-item demand.

Finally, the quality of American goods has suffered in recent years. In fact, since World War II the relative positions of the United States and Japan have virtually transposed themselves in terms of quality of goods produced. This situation became especially apparent in the automotive industry, but is now evident almost everywhere. According to Phillip Crosby, in his book *Quality is Free*, it is not unusual for United States manufacturers to spend 15% to 20% of every sales dollar on quality costs associated with prevention, appraisal, and failure. While this is true, quality costs should generally run no higher than 2.5% of sales. When we look at the costs of quality and their historical impact, it is easy to understand why we produced 40% of the world's motor vehicles in 1960, but dropped to 20% by 1980 while Japan went from 3% to 28% in the same period. The postwar growth of American industry was oriented toward high volumes, specialization, and technological advances. Foreign competition, especially the Japanese sector led by Dr. W. Edwards Demming, concentrated

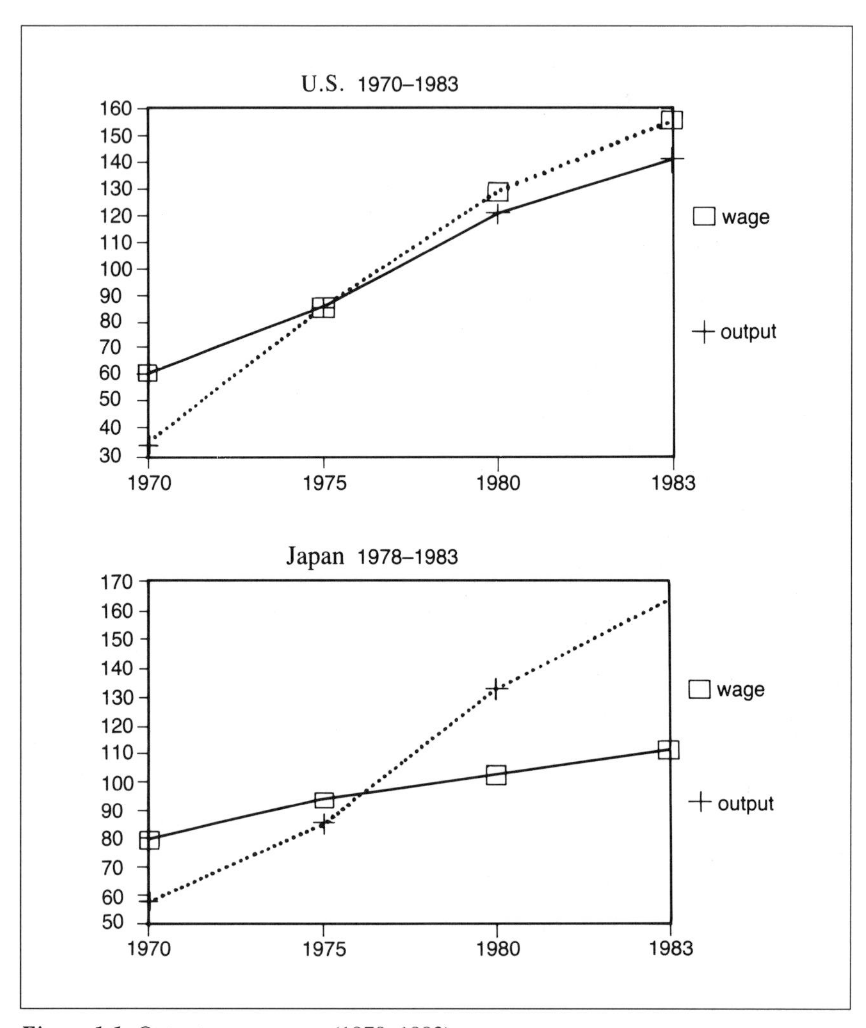

Figure 1-1. Output versus wage (1970–1983).

on shipping high quality goods, and constant improvement through Statistical Quality Control (SQC). By the 1970s, Japan had moved beyond the focus of SQC to encompass all of Deming's 14 points, and initiate a program known as Total Quality Control (TQC).

As industry after industry falls to foreign competition, America's trade deficit continues to widen. (See *Figure 1-3*). Cameras, video recorders, audio equipment, machine tools, and steel have become almost extinct in terms of domestic production. The goals of American industry in overcoming this situation appear, at least on the surface, to be contradictory. We are confronted with the need to

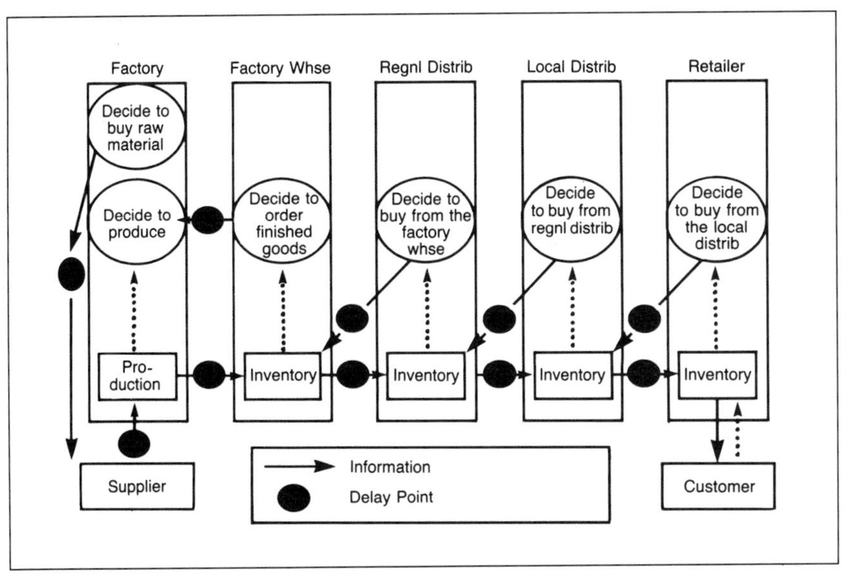

Figure 1-2. Typical U.S. production and distribution.

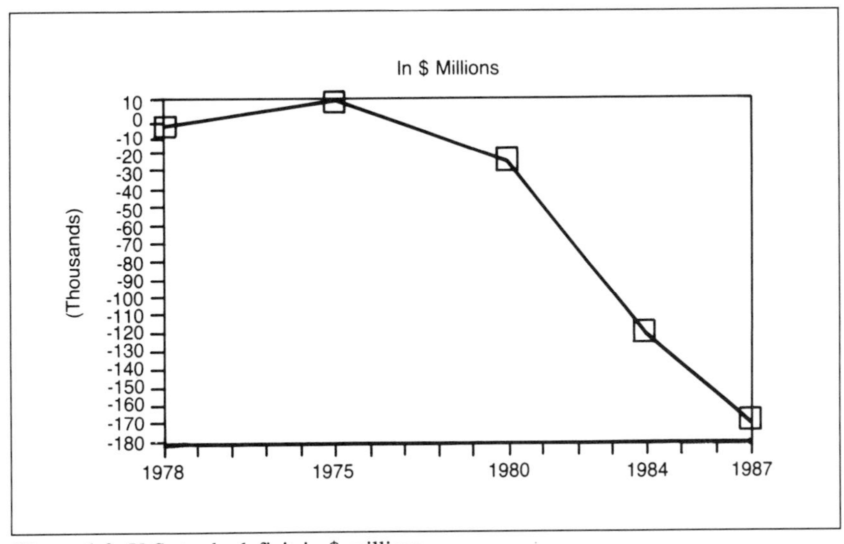

Figure 1-3. U.S. trade deficit in $ millions.

make exactly what is needed (not more, not less, and not the wrong things), with greater efficiency and less cost, and at higher quality levels. We must simultaneously reduce throughput times from design delivery, eliminate defects, and reduce product cost.

The path to regaining manufacturing excellence, and achieving these goals may be viewed as a five phase maturation process. The major steps in this process include:

Phase I

Phase I is the most primitive phase of development, characterized by a general struggle with fundamental controls. Recognition of the need for controls is indicative of the beginning of this phase. By the end, critical data has been identified, documented, and is maintained. The importance of managing most of the critical resources of the business is recognized. (Critical resources here include capacity, inventory, human resources, equipment, and suppliers among others).

In the area of quality, there is typically no recognition on the part of management that a problem exists. Even when the need for accurate bills of material, inventory data, etc. is recognized, there is often no real understanding of quality-related costs. There is no tracking of warranty costs, inspection costs, rework costs, or aggregation of these costs into a cost of quality figure. Scrap costs are maintained by some companies in this phase, but even they are generally pure material values with no accurate labor and overhead additive. By the end of Phase I, defects may be noted by inspectors on a part number by part number basis, but only exceptionally problematic situations are dealt with.

In terms of layout, the manufacturing facility in a Phase I environment is usually a mess. Housekeeping is atrocious with dunnage, trash, and offal lying everywhere. Lighting is often poor, as is the noise level and general visibility of operations, and additional equipment is put wherever there is room. Raw and work-in-process (WIP) material is stored haphazardly with marginal tracking. Production occurs in ''waves'' through the factory, with excessive setup/changeover times and negligible work load balancing between operations.

Production and Inventory Management in a Phase I setting is usually composed of a reorder point system which is based on questionable information. Since inventory data, bill of material integrity, and routings are usually poor, there is little foundation on which to build. Coupling this situation with equally poor production, scrap, and part substitution reporting, yields an almost hopeless control picture. Because of this lack of tools and understanding, line supervision becomes frustrated with the poor prioritizing and launching of orders. Continuous expediting is required, and line supervisors begin to reprioritize and control production on the shop floor themselves, responding to shipping requirements with ''Hot'' list meetings and overtime.

Procurement in the Phase I environment is typically a tense activity, filled with anxiety from wildly fluctuating demand and poor ongoing controls. Since little long-range visibility exists, long-range contracting is all but impossible. In addition, the daily production expediting is usually translated into similar activities for purchased goods. Even good supplier relationships are strained under these conditions, and most of them quickly become adversarial. Pressure to minimize the price on each part and raw material purchase aggravates the situation. Quality, delivery, and price all suffer in order to gain responsiveness.

Purchasing is a hive of activity, but most of it is firefighting. There is little opportunity for intelligent buying, and little understanding of the need for supplier visits and performance monitoring.

Design Engineering in Phase I is an effort to achieve low cost and fitness for use. Little is known about the changing needs of the customer, or how current designs affect product performance. The only manufacturing involvement in the design process involves urgent requests from the shop floor when the design cannot be produced, will not fit in assembly, or must be changed due to part shortages. Toward the end of Phase I, there may be a print review process for new designs, where purchasing and mechanical engineering have an opportunity to look at prints before the new design is formally adopted.

Support areas in the Phase I environment are characterized by adversarial relationships with manufacturing, and with each other. As a result, there is often little or no visibility between the disciplines. Goals and objectives are seldom complementary, or shared. Frequently, no common performance measures exist. In terms of cost accounting, for example, there is frequently no way to accurately cost the end product. Production standards and labor valuation do not exist, or are inadequate. In addition, bills of material are not reliable, so "rolling up" costs based on their integrity is of little value.

There is little to be said of employee involvement in the Phase I environment. There is generally no perceived need for the ideas of direct labor or staff personnel. If there is a program, it usually consists of a suggestion box. Good ideas are ignored or rejected often enough to kill any enthusiasm that may have existed, and lip service is all that remains of management support.

In summary, then, Phase I involves the installation and maintenance of fundamental controls. In this stage, bills of material, routings, inventory control, labor standards, and production reporting are installed. Only when these practices are in place does an adequate foundation exist for Manufacturing Resource Planning in Phase II.

Phase II

The Phase II manufacturing environment involves recognition by management that measurement and control of all resources is essential. The critical nature of timely, accurate information flows usually comes to light as well. There is often a real underlying perception developed during this period that adequate data levels can solve almost any problem. Along with the burgeoning software activity, the functions of manufacturing management usually become more complicated and specialized.

Organization levels swell, and functions like design engineering, production planning, purchasing, and sales become distanced from the manufacturing process. Visibility of the operations diminish, and production controls become more theoretically sound but less realistic.

In the quality area, management recognizes that there is a significant impact from poor quality on profitability levels, and begins to establish measurement devices. The measurements typically implemented here are neither comprehensive nor accurate. They generally center around incoming inspection of material, sample inspection of work in process, and final inspection of finished goods. Little is done with the data that is gathered, aside from using it to berate manufacturing line supervisors. Responding to quality problems entails increasing inspection frequency, increasing scrap allowances, and forcing the employee to perform any required rework. Occasionally a manufacturer in Phase II will have incorporated Statistical Process Control (SPC), but in these situations little is done with the charts. Reported costs of quality will be 2% to 4% of sales, while actual costs run anywhere from 15% to 25%.

In terms of layout, traditional manufacturers in the Phase II environment tend to compartmentalize their operations, orienting layout toward function. Metal stamping will be in one department, machining in another department, assembly in still another. In this situation, product flows are stretched in terms of both time and distance, crisscrossing paths as they move back and forth between departments and WIP stores. Work centers are organized into rows or banks with plenty of room for large interoperation queues and material movement.

Production and Inventory Management in a Phase II environment is dominated by material requirements planning (MRP), and manufacturing resource planning (MRPII). The planning functions supported by MRP and MRPII software have proven successful in thousands of American manufacturing organizations. While this is true, the levels of success vary greatly from installation to installation. These systems involve substantial investments in computer hardware and software, training, and maintenance. Their primary purpose is planning of resource utilization, based on periodic runs of a production plan and/or master production schedule. Production orders are generated and launched based on due dates, and in "economic order quantities" (EOQs). Work centers are scheduled independently, based on earliest due date of available order for each work center, as capacity opens up.

There is probably no better planning tool for manufacturing than MRP. However, as an execution device, the shop floor scheduling and controls portion of MRP/MRPII systems are deficient. Because of the "Push" orientation of MRP order launch, and constant reprioritization through expediting and de-expediting, buffers of raw, WIP, and finished goods inventories are utilized to compensate for lack of coordination in production. As a result, inventory levels and lead times are increased while cusstomer service levels are diminished.

Procurement in a Phase II environment reflects the awakening of management to the need for some level of cooperation with suppliers. This frequently originates in VA/VE activities (discussed later), or as a result of management mandate to reduce raw material costs. However it may originate, the result is that a buyer and one or more suppliers working together to improve cost, delivery performance, and/or quality levels on purchased parts and materials. At this stage, improvement activities are typically oriented toward individual part numbers. The efforts are seldom part of a formal improvement plan, and are not well monitored or directed. Even though relationships are strengthened, this activity is constrained by an unwillingness to share pertinent data (costs, production volumes, etc.) between suppliers and buyers, and by a multiple-source philosophy.

Generally speaking, then, procurement in this phase is part number oriented, and price and delivery driven. Improvement efforts exist, but they are narrowly focused and not well managed.

Design Engineering in Phase II begins to evolve, as management recognizes the need for involvement by manufacturing in the design process. Spurred by the need to minimize costs, value analysis and value engineering programs are frequently initiated. Make vs. buy analyses become a formalized activity, further contributing to interdisciplinary cooperation and communication between Design Engineering, Purchasing, and Manufacturing.

Another development in Design Engineering which is common to Phase II is the adoption of computer aided design (CAD). Associated with this new tool is some recognition of the need for standardization of components, but this effort typically begins with low-value components like threaded fasteners. Finally, the tendency to "systemize" often spills over into the area of Engineering Change Notices (ECNs). ECNs are often tracked as through they were production orders, with frequent expediting.

Support areas in the Phase II environment are characterized by learning to collect and manage a great deal of new information. Associated with MRPII are capacity requirements planning (CRP), distribution requirements planning (DRP), and a number of other systems. Each of these modules is impacted by the others, and provide volumes of information which must be analyzed and acted on at least weekly. Managing the information, and managing operations based on the information, involves a substantial learning curve. Each area needs to develop an understanding of what other departments do, in order to manage their inputs and outputs effectively. At the same time, the enormous volumes of data which must be reviewed, analyzed, and acted on drive individuals toward a very focused perspective of the business. It becomes very hard to see the forest for the trees.

A byproduct of all of this mechanized and structured planning is often a real mind-set orientation toward the planning function. Execution of the plan is de-emphasized, and frequently planning grows into a multilayered, labor-intensive process.

Formal budgets and business plans are developed for each layer of the organization. As forecasts and markets change, the plans must be revamped. Eventually, planning activities like most others evolve to the point that 20% of the activity accounts for at least 80% of the value.

Employee involvement in Phase II is still very limited.

There is often an employee suggestion program, consisting of a suggestion box, and a committee which reviews suggestions and decides on their viability. In these programs, individuals responsible for evaluating the merits of the suggestions are usually the very individuals who are responsible for the areas involved. As a result, many suggestions which are sound, money saving ideas are rejected out-of-hand because of the defensiveness or embarrassment of reviewers. Immediate reactions are along the lines of "that would never work because...." A better answer would certainly be "how can we make this work?"

Mistrust and suspicion are fostered in this environment, and suggestions level off or decline in quantity, quality and percent of implementation.

In summary, Phase II involves the implementation and use of sophisticated planning and control systems. Information processing is emphasized, and often becomes as much an objective as it is a tool. Planning is the central focus of activities, and execution often suffers.

Phase III

Phase III is the period in which management recognizes that a shift in emphasis must be undertaken from a planning orientation toward one of execution. Visibility and resolution of problems are underlying themes in this environment, as is the utilization of the companies' most valuable resource—its people.

In the area of quality, management usually becomes aware in this period that simply measuring defects and/or process capability will not insure improvements. As problems are identified by inspection, and out-of-control processes become apparent through statistical process control (SPC), problems are analyzed for their underlying causes, and resolved. Costs of quality are more accurately defined, but not correct. Typically, reported COQ is 8-10%, with actual costs of roughly 12%.

In terms of layout, management in the Phase III environment recognizes the value of orienting manufacturing equipment around product flow rather than function. Machining departments, welding departments, and assembly departments give way to manufacturing cells containing machining, welding and assembly operations. Work load within and between cells is balanced, to minimize interoperation queues, and provide manpower flexibility. Product travel distances are minimized, as are nonvalue-adding activities.

Production and Inventory Management in the Phase III environment usually involves less emphasis on launching manufacturing orders, and more on "pulling" production via a final assembly schedule. Through reduced throughput and setup times, more flexibility is available to meet schedule changes. Since individual operations are "connected" via cellular layouts and pull systems, individual work center scheduling is greatly diminished. Utilization of pull signals (or Kanbans) also greatly reduces the need for expediting and de-expediting. With reductions in defect levels and setup times, the need for constantly recalculating EOQs and scrap allowances is eliminated. To summarize, production scheduling is largely reduced to final assembly scheduling.

Procurement in the Phase III environment involves the establishment of supplier relationships which are akin to partnerships. Purchasing management begins to revolve around commodities, as opposed to individual part numbers. More data is shared with suppliers, and fewer suppliers are used. Ongoing vendor review and certification programs are established. Contracts are expanded to encompass quality and delivery criteria, as well as long-term cost reduction. Supplier capacity is purchased, and suppliers begin to be treated in many ways as though they are an extension of the manufacturer's own company.

Design Engineering in Phase III is also impacted by the purchasing "partnership" concept, as suppliers become involved in the manufacturer's design activities. Other changes in design engineering which frequently emerge at this stage include Early Manufacturing Involvement (EMI), and Current Product Review (CPR). The underlying theme is involving manufacturing in the design process. Also indigenous to Phase III design engineering is the concept of part management by family. Often, this process is "systemized", in a Group Technology (GT) program. GT involves coding part characteristics into the computer, such that it becomes easier to standarize part usage by looking up existing components during the design process. For example, one might logically group gears, with bevel-tooth gears as a sub-family, 32-tooth as the next level, specific inner dimensions and outer dimensions at the next levels, etc.

Support areas in the Phase III environment like production and inventory management described earlier, shift their focus from planning to execution. In addition, goal commonality between disciplines is enhanced significantly. Marketing, for instance, is able to take advantage of newly reduced throughput times to boast of responsiveness, and enhanced quality levels to entice potential customers. Cost Accounting concentrates less on WIP inventory valuation (since there is less WIP to value), and traditional product costing is frequently converted to some form of process costing. Distribution professionals, recognizing the merits of eliminating nonvalue-adding activities and other forms of waste, look beyond distribution requirements planning (DRP) to compressed delivery times and maximized efficiency. The information systems department is not required to process as much information to support shop floor control (SFC) activities, but the data that is needed is required more quickly. Information will be aggregated and managed in different ways, as well.

Employee involvement in Phase III begins to emerge, and develop into a driving force for continuous improvement. A formal EI program is set in motion to harness the constructive energies and ideas of the organization's most underutilized resource—its people. Training is provided in problem solving skills, and employees are organized into problem solving work groups, with a formal reporting structure and budgetary resources. Initial problems are assigned, and by the end of Phase III, problem solving activities have begun.

In summary, Phase III is predominated by the Just-In-Time philosophy of waste elimination and continuous improvement. It is typified by interdisciplinary communication, goal commonality, and a shift in emphasis from planning to execution.

Phase IV

Phase IV is typified by the returning shift in emphasis from simplified, synchronous production to automation, experimentation, and technological advancement. Redundant or dangerous tasks are performed via robots. Materials are transported between areas by automated devices. Electronic data interchange, bar coding, numerically controlled (NC) equipment, computer aided manufacturing (CAM), and a number of other technological solutions become the focal point of management attention.

In the area of quality, Phase IV activities are firmly centered around prevention (as opposed to detection) of defects. Beyond the use of control tools (statistical process control charts, Pareto analysis, cause-and-effect diagrams, data stratification, check sheets, histograms, and scatter diagram), quality function deployment (QFD) and designed experiments are utilized to get the product and process right from their outset—at design. This activity is predicated on the belief that it is most beneficial to improve product process and design before manufacture. In the Phase IV environment, the definition of quality is expanded beyond "conformance to specifications" to include concepts such as timeliness and cost. Reported cost of quality is typically about 6.5% and actual costs are roughly 8%.

In terms of layout, the Phase IV Environment is still very much typified by process oriented material flows, with many of the material handling activities eliminated via Phase III JIT implementation, and a large percentage of the remaining handling performed by automated devices such as Automated Guided Vehicles (AGVs). Also indigenous to Phase IV is the tendency toward reduced numbers of operations, with more being done in fewer steps through improvements in product and process design.

Production and Inventory Management in the Phase IV environment is much less oriented toward shop floor control, due largely to the implementation of pull systems in Phase III. Production is prioritized and driven from a final assembly schedule, rather than individual work center scheduling. In addition, individual component inventory consumption and production reporting are often handled automatically by "back flushing" bills of material at the end of the assembly process. This procedure is often further enhanced by the use of bar coding.

Procurement in the Phase IV setting cements the partnerships initiated with suppliers in Phase III by driving toward single sources for purchased goods, and emphasizing shared data with suppliers.

Contracts are negotiated for supplier capacity over multiple years, with releases for shipment visible well in advance to vendors. Supplier certification is emphasized, and suppliers are regularly included in the relevant design reviews.

Design engineering in Phase IV involves a heightened use of computer aided manufacturing (CAM) to develop manufacturing routings directly from designs, based on part characteristics. In some cases, these activities are carried even further, so that numerical control (NC) tapes are generated, or machines are programmed via computer numerical control (CNC). Building quality into the design is paramount, and minimizing tools, equipment, and operations in the manufacturing process is also emphasized.

In terms of support areas, Phase IV is a picture of adjustment, compensating for the substantial changes in manufacturing processes invoked by Phase IV. Marketing is able to capitalize on newly shortened manufacturing throughput times as a sales tool. In an effort to shorten the order entry cycle time to match manufacturing, remote order entry terminals are often issued to the field sales force. Accounting, after coming to grips with the changes in Cost Accounting associated with JIT, will find that Phase IV requires significant efforts in the area of equipment justification. Determining return-on-investment (ROI) for AGVs and other equipment is an ongoing process in most of Phase IV.

Employee involvement in Phase IV is a mature, operational organization. At this stage of development, management has come to recognize the constructive power of EI, and utilizes it regularly to address difficult operational problems. EI comprises the bulk of the ongoing improvement activities which are residual from the JIT program.

In summary, Phase IV can be described as a return to technology oriented focus, with continuing adjustment on the part of support areas to the streamlined operations resulting from Phase III. Much of the transitional work associated with Phase IV revolves around the justification and implementation of automation in the factory.

Phase V

Phase V, the most advanced documentable phase at this point, is predominated by information integration activities. From supplier through customer, and from CEO to factory worker, critical information must be accumulated, analyzed, and distributed to the appropriate parties. Coordinated, timely responses to technological, market, environmental, and competitor changes are essential to achieve world-class manufacturing status.

In the area of quality, COQ is reduced to roughly 2 or 3%, both reported and actual. Process variance is usually under firm control, and design is heavily influenced by manufacturing and marketing. The entire organization has come to recognize that it has a real responsibility for quality. It is uppermost in the minds of buyers, manufacturing supervisors, design engineers, and top management executives. Customer needs and competitor abilities are regularly benchmarked, and quality is continuously redefined and reemphasized. Quality goals are then set and reset based on the findings of this benchmarking process so that the driving force behind organizational improvement is customer needs and the desire to outperform the competition.

In terms of manufacturing layout, Phase V is typically very much like a process industry flow, even in a discrete manufacturing environment. Automation has replaced most manual labor, and material flow is virtually continuous from raw material to the customer, with minimal wasted travel distance and processing time. Changeover times are reduced to virtually zero via flexible manufacturing systems and Programmable Logic Controllers (PLCs), and are generally handled remotely through the initiation of computer program changes.

Production and Inventory Management in Phase V is largely comprised of master production scheduling, and final assembly schedule maintenance. The flow-oriented nature of production ensures that individual work center priorities remain intact based on final assembly schedules. Little or no expediting/de-expediting is required, since end-to-end production times have been minimized. Virtual lot sizes of one combined with balanced work loading make close-in schedule changes no more difficult than the timely procurement of raw materials.

Procurement in the Phase V environment is a picture of cooperative and continuous improvement effort between the manufacturer and suppliers. A great deal of data is shared between the manufacturer and suppliers, including production schedules, cost and profit levels and designs still on the drawing board. Purchase order releases against multiyear contracts, advance shipping notices, acknowledgements of receipt, and funds are all transmitted between manufacturer and supplier electronically. Joint manufacturer and supplier improvement teams and problem solving teams are continuously organized to attack problems, then disbanded as they accomplish their objectives.

Phase V design engineering involves continuous input in the forms of field performance data which is gathered and communicated via E.D.I., manufacturing involvement program suggestions, market analysis data, and competitive capability and design intelligence. The design itself moves from CAD workstations into databases from which it can be retrieved in digitized form and converted automatically into manufacturing routings. The routings can then be converted to computer programs with the aid of group technology that will produce component parts, sub-assemblies, and final assemblies via CNC equipment on the factory floor. Bills of material and/or bills of resources are also generated for use by Production Control and Purchasing automatically, as a result of this process.

In support areas, Phase V is typified by immediate information feedback between disciplines and management levels. Data flow is more clearly defined, and structured, but is also more dynamic. Since the entire operation is oriented toward and managed by information flows, there is a strong return to emphasis on data integrity, and Management Information Systems gains much more prominence in the organization. Some companies establish high level management positions such as CIO (Chief Information Officer) in these environments.

Employee involvement in Phase V has evolved into an integral part of operations management. Continuous improvement activities reach into every part of the business from supplier operations to customer service, with employee improvement and problem solving teams continuously redefining their objectives and project priorities. EI in Phase V becomes so valuable a part of the improvement process, that management of the business without it seems incredulous.

Maturation Process

The maturation process may be visualized, then, within the context of the matrix shown in *Figure 1-4*.

This particular path has evolved from the development of increased capability and sophistication. In fact, with one notable exception, when a business moves beyond the point of sound basic controls, the computer becomes more and more essential and integral to improvement. Along with the increase in sophistication, generally there is an associated increase in cost. The further into this process a company travels, the more substantial that cost becomes. Again, there is one notable exception. The exception is Just-In-Time. Just-In-Time stresses simplification rather than sophistication, and is generally far less expensive a program to undertake than MRP II, FMS, CIM, etc. In addition, the benefits from Just-In-Time programs are frequently much greater than the other phases. In my

	Phase I	Phase II	Phase III	Phase IV	Phase V
QUALITY	COQ: Unknown but 20% — No recognition that a problem exists.	Reported: 3% Actual: 18% — Recognize, begin to measure defect rates.	Reported: 8% Actual: 12% — Identification and resolution of problems.	Reported: 6.5% Actual: 8% — Prevention Variation Research.	Reported: 2.5% Actual: 2.5% — Virtually zero defects, no variance.
LAYOUT	Based on available room for new equipment.	Functional layout with WIP stores.	Process oriented flow with diminishing WIP, movement, cellular mfg.	Process orientation with diminishing handling and numbers of operations.	Continuous flow, raw to finished. Virtually no human intervention.
P&IM	Basic Controls -BOM -Inv. -Routing (ROP/min–max).	Requirements Planning (MRP, MRP II) Master Scheduling.	Just-In-Time Pull systems, etc. More flexibility required.	Automation Final assembly schedule driven backflush thru bar coding.	CIM Complete data sharing, control by data.
PROCURE	Adversarial, part # oriented, price driven.	Cooperative, part # oriented, price and delivery driven.	Partnership, commodity driven, quality and delivery oriented.	Partnership, commodity and capacity driven, single source oriented.	Partnership, continuous improvement driven, data sharing oriented (EDI).
DESIGN	Fit for use No non-engineering involvement except print review.	VA/VE Make/buy analysis CAD standardization ECN control.	(EMI/CPR) Manufac-turability G.T.	N/C equipment CAM.	PLCs, design from eng thru BOM to PLC.
SUPPORT	Adversarial w/mfg., each other. No interdisciplinary visibility.	Visibility of production data, formal planning as teams. Beginning of integration and more dynamic data.	Less emphasis on planning, more on execution. Inter-disciplinary teams. Common goals.	Common performance measurements. Formal, structured justifications.	Immediate info feedback. More structured but dynamic. Emphasis on info management.
EI	Adversarial No recognition of need.	Light need recognition. Mistrust. Emp. suggestion program.	Formal EI process initiated. Training begun.	Ongoing EI program with formal structure and management.	Continuous improvement.

Figure 1-4. Manufacturing Development Spectrum.

own experience, I have found that it is not at all uncommon to experience these kinds of benefits with proper JIT implementations:

WIP inventory reductions of up to 99%
Throughput time reductions of up to 91%
Productivity increases of more than 30%
On-time shipping performance improvements of 58%
Procurement lead time reductions of 98%
Purchased part cost reductions of more than 20%
Quality cost reductions of up to 50%
Design change throughput time reductions of more than 50%

Because there is little software or hardware involved, costs are much less than the other programs listed. While costs and benefits vary with every implementation, it is easy to see why Just-In-Time is such an appealing philosophy. With enormous potential benefits and comparatively low cost, with attractive underlying principles like simplicity and common sense, and with a unique ability to meet the seemingly contradictory goals of current American manufacturing, what could possibly stand in the way of its wide-spread adoption?

Part of the answer is merely a question of time. A significant segment of our manufacturing population is still struggling through the Basic Controls/MRP/MRP II area (see *Figure 1-5*). A second factor is lack of awareness. American manufacturers are largely unaware of the benefits associated with JIT implementations, and their universal application. Lastly, American manufacturers who are aware of Just-In-Time, and what it can do for them usually do not know how to go about implementing it.

With the disappearance of one American industry after another, it has become obvious that time is a luxury we can no longer afford. This book is a modest effort to provide some of the required awareness and education to accelerate the recovery process.

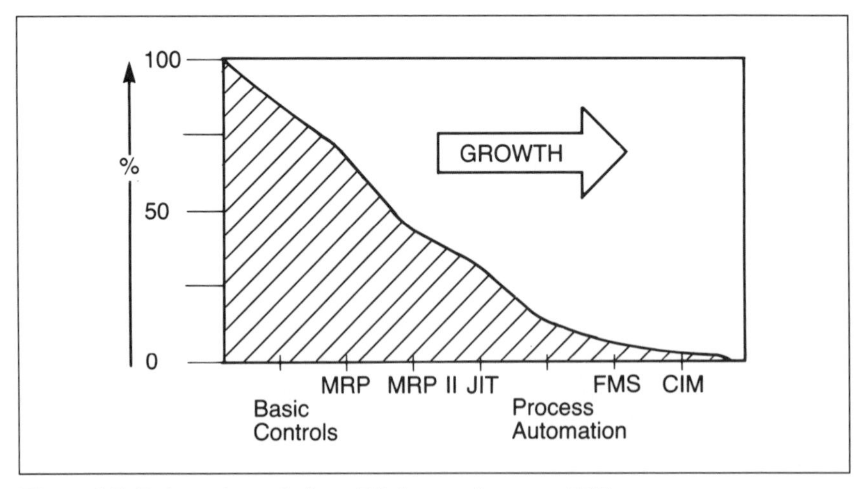

Figure 1-5. Estimated population of U.S. manufacturers 1987.

The reader will find that this book is easily regarded as two sections: Section I deals with background material, definitions, and an explanation of fundamental Just-In-Time concepts. Section II is a kind of "cook book" description of the implementation of Just-In-Time in a manufacturing company. I have seen this kind of program work very well (in whole and in part) in high tech, high volume operations as well as very low tech environments. It is easily modified, and modular.

A DEFINITION OF JUST-IN-TIME

2

Just-In-Time is a philosophy which has as its objective the elimination of waste. The philosophy applies in all kinds of industries, including banks, hospitals, services, and manufacturing.

In manufacturing environments, waste may appear in many forms, including defective parts, excess inventory, unnecessary material handling, setup, and changeover times, to name a few. By concentrating on these areas, Just-In-Time programs have racked up impressive results. Some better known examples of American applications include:

- Black and Decker, who has reported that, with the application of Just-In-Time concepts and a great deal of hard work, a pilot plant increased volume of throughput by 300%, increased inventory turns from 4 to 40, and reduced manufacturing lead times by more than 50%.[1]
- Omark Industries, who has reported that their version of JIT, called ZIPS (Zero Inventory Production Systems) reduced 6 1/2 hour setup times to one minute and forty seconds, cut space requirements by 40%, reduced lead time from 21 to 3 days, and lowered plant-wide inventories by 50%.[2] All of these accomplishments notwithstanding, Omark also reported 10% productivity improvements, 97% order fill rates, and 35% reductions in manufacturing costs.
- General Electric studied and applied Just-In-Time manufacturing techniques in a factory producing dishwashers with the following results: nine miles of conventional conveyor replaced by 2.9 miles of computer-controlled nonsynchronous conveyors and robot handling systems; human material handling reduced from 28 occurrences to 1; internal failure cost reductions of 51%; inventory turns increased from 12 to 34; inventory investment reductions from $9.7 million to $3.7 million; throughput time reductions from 5 or 6 days to 18 hours; plant output increased by 20%. All of this was achieved with a 40% reduction in required floor space.[3]
- Harley-Davidson has attributed their very existence to the application of JIT, which they refer to as their MAN (Material As Needed) program. With MAN in place, Harley-Davidson was able to reduce their break-even point 32% in 1982. Between 1982 and 1984, raw material and work in process inventory turns increased from 6 to 17, and by 1985 were around 20. Setup time reductions of 75% have been realized, and assembly throughput has gone from 3 days to half a day. Direct, indirect, and salaried employee productivity improved 37%. Supplier bases were consolidated by 23%, internal scrap costs were reduced 52%, and rework costs dropped by 80%. Defects per unit are reported to have dropped by 53%, and warranty costs per unit are down 46%.[4]

While the Just-In-Time philosophy of waste elimination proves helpful in virtually all types of manufacturing (and service) environments, some types of manufacturing offer more opportunity than others.

Generally, process industries (paper mills, chemical manufacturers and food processors) pose a greater challenge in terms of setup time reduction and process flow improvemtents because of the "connected" nature of their operations. Repetitive and discrete manufacturers (metal stamping and machining) tend to encompass more opportunity in these areas because of their predisposition toward functional area layouts, and extensive/frequent machine setups. Conversely, process industries generally have a great deal of opportunity in the areas connected with materials planning and procurement. These industries often use relatively large quantities of raw materials, and may have significant advantages in leverage to negotiate supplier improvements such as delivery, quality, and cost.

Even within an industry, opportunity levels will vary from company to company because of skill levels, layout styles, organization structures, labor union constraints, management abilities, geographic location, financial resources, and several other factors. The best way to evaluate opportunities in a specific manufacturing environment is to perform an overall Opportunity Asssessment. This procedure will be discussed later. The important point to remember is that opportunity areas vary in size depending on the industry involved, and may vary widely within industries depending on the characteristics of the specific company.

Waste elimination and continuous improvement are increasingly appropiate in the office environments surrounding manufacturing operations. Purging paperwork processes of duplicated effort and otherwise inefficient activities has become a critical aspect of reducing costs, and JIT principles can be effectively applied in these situations.

Just-In-Time may also be utilized in the general management arena. It is not uncommon to find waste elimination techniques to be useful in streamlining product lines, product options offered, markets and industries served, and capital investments in facilities/equipment.

Before we discuss specific components of Just-In-Time as applied to manufacturing, it will be advantageous to identify a few underlying principles of JIT. These include:

1. The ***ONGOING*** nature of a JIT program. It is extremely important to understand that Just-In-Time is not a project, because it has no end. Once JIT improvement activities are under way, they should continue indefinitely. The environment should be transformed to one of continuous ongoing improvement, and cooperative management/labor endeavor.
2. The benefits of ***SYNCHRONIZATION*** or balance. This process involves the matching of throughput times from operation to operation during the course of manufacturing and support functions, so all production occurs at a common rate, or "drum beat".
3. ***SIMPLICITY***; the view that simpler is better. A continuous effort is made to perform required operations with fewer resources (time, personnel, and equipment) and in a less complicated fashion.

With these basic tenets in mind, let us turn our attention to the individual modules of JIT.

References

1. Albert G. Wordsworth, ''Survival,'' Strategic and Tactical Issues in Just-In-Time Manuacturing, 1985 Conference Proceedings (Wheeling, IL: Association for Manufacturing Excellence, Inc.).
2. Michael J. Rowney, ''Omark Industries' Zero Inventories Production System (ZIPS),'' Strategic and Tactical Issues in Just-In-Time Manufacturing, 1985 Conference Proceedings (Wheeling, IL: Association for Manufacturing Excellence, Inc.).
3. Raymond L. Rissler, ''The General Electric Dishwasher Plant at Louisville, Kentucky,'' Strategic and Tactical Issues in Just-In-Time Manufacturing, 1985 Conference Proceedings (Wheeling, IL: Association for Manufacturing Excellence, Inc.).
4. Thomas A. Gelb, ''The Material as Needed Program at Harley-Davidson,'' Strategic and Tactical Issues in Just-In-Time Manufacturing, 1985 Conference Proceedings (Wheeling, IL: Association for Manufacturing Excellence, Inc.).

THE COMPONENTS OF JUST-IN-TIME

3

When the Just-In-Time (JIT) philosophy is used to construct a program for manufacturing improvement, it should always contain eight primary components. These are:

- Organization of the program
- Quality
- Simplified, Synchronous Production
- Process-Oriented Flow
- Advanced Procurement Technology
- Improved Design Methods
- Enhanced Support Functions
- Employee Involvement

Organization

A large percentage of the "failures" associated with Just-In-Time implementations are attributable (in part or in whole) to inadequate planning an organization. The first step in achieving a sound, comprehensive plan for implementing Just-In-Time in manufacturing then, involves organizing the other seven components into a cohesive whole, with each element representing a facet of the JIT "emerald" shown in *Figure 3-1*.

The actual process of implementing a JIT program into the manufacturing environment in turn requires eight implementation steps, which can be visualized using the following model in *Figure 3-2*.

1. Program Organization is comprised of awareness and education sessions and the organization of the steering committee, assessment team, and

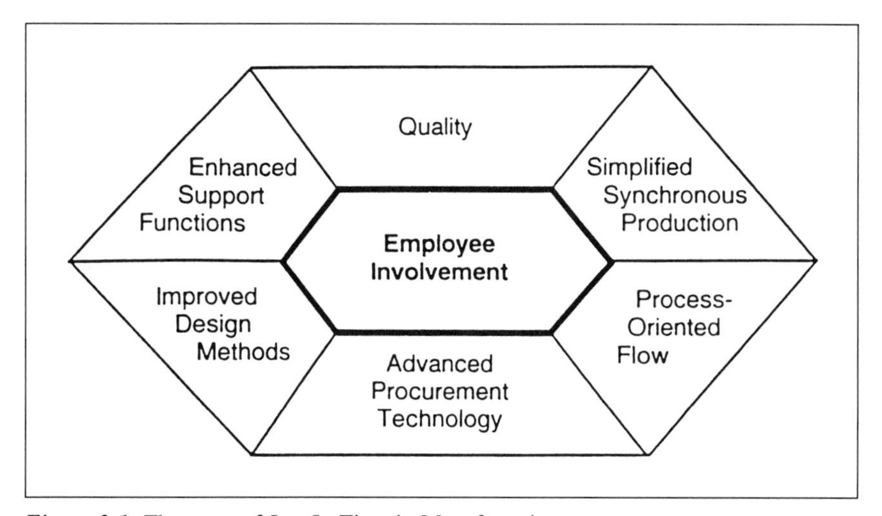

Figure 3-1. The steps of Just-In-Time in Manufacturing.

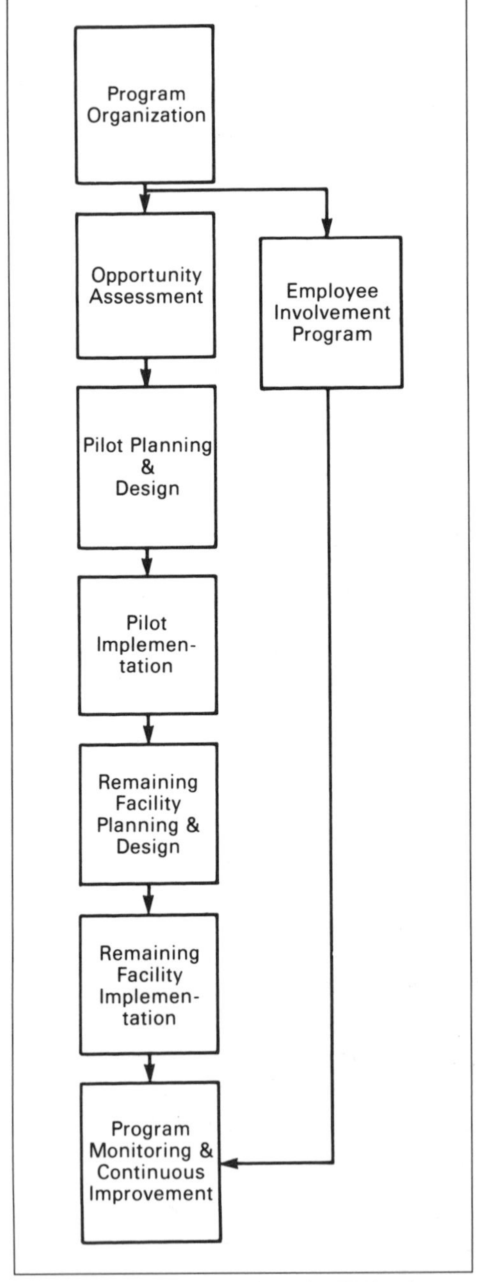

Figure 3-2. The organization of a JIT implementation program.

employee involvement program. During this step, a strategic business assessment is performed, so that subsequent opportunity assessment and design activities appropriately account for projected business developments.

2. The Opportunity Assessment entails a series of activities designed to identify opportunity levels by area and by product line. The data gathered during this step will allow improvement activities to be properly prioritized and realistic goals to be set. At the conclusion of this step, employee involvement (EI) activities are initiated, and pilot implementation projects are identified.
3. Pilot Planning and Design involves the organization and training of pilot project teams. Pilot floor layouts and other changes are designed, and implementation plans are developed for each project. Pilot performance measures are also defined.
4. Pilot Implementation is the activation and monitoring of pilot projects. Problem-solving activities are initiated from the pilot projects, and improvements are reported to management.
5. Remaining Facility Planning and Design is comprised of reviewing what was learned from pilot projects and applying that knowledge to develop a sound conceptual design of the remaining facility. Specific implementation goals

and projects are identified, along with appropriate performance measures.

6. Remaining Facility Implementation covers remaining facility implementation activities (equipment movement, organization changes, etc.). Specialized training is provided as required, and implementation progress is periodically reported to management.
7. Program Monitoring and Continuous Improvement is initiated during the remaining facility conversion activities. Overall program progress is documented, and the program is guided through its transition from discrete implementation projects to continuous problem solving and improvement. Program administration is transferred to the employee involvement structure, as remaining facility implementation projects are completed.
8. Employee Involvement activities are initiated in parallel with the other JIT program modules, and eventually they become the heart of the entire waste elimination process. Even beyond remaining facility conversion, continuous improvement will be needed. Problems arise to be resolved, and variances will need to be reduced.

Quality

The Just-In-Time philosophy is best supported by a Total Quality Control (TQC) program. This program continuously seeks to eliminate defects—first by identifying and removing existing defects, and finally by preventing defects before they can occur. In order to accomplish this, critical quality characteristics must be defined.

The definition of quality should revolve around the customer's perception of quality. ''Customer'' here includes both subsequent internal operations and the ultimate consumer of the product. When the definition of quality is established, steps can be taken to:

- Ensure that designs will produce quality products.
- Ensure that current or forthcoming process capability exists to produce the design.
- Ensure that appropriate quality levels are available on a consistent basis in raw materials and purchased parts.
- Ensure that appropriate measurement devices exist to monitor all defined quality levels.
- Ensure that appropriate measurement devices exist to monitor and improve all defined process capabilities. Techniques utilized here include Statistical Process Control (SPC) and Variation Research.

Once these assurances exist, employee training activities can be undertaken. All employees must come to regard quality as their responsibility and learn to:

1. Audit previous operation quality levels, and
2. Continuously insure the quality of their own processes and products. This amounts to a concept called ''quality at the source,'' which is an extremely valuable program in its own right. The objective is to prevent defects at each step in the process, rather than trying to inspect for them at a later time.

Simplified Synchronous Production

Simplified Synchronous Production (SSP) is a program of converting individual activities into aspects of a continuous flow that is synchronized to end-product demand.

Utilizing SSP, production rates are matched as closely as possible to actual end-product sales. This can be done as lot sizes are brought to minimal levels through reduced setup and changeover times. Moreover, production rates must be matched from work center to work center within the production process. In this manner, production occurs as though following a common drum beat.

Simplified synchronous production also involves housekeeping issues that support simplified, safe production. Visibility and accountability for all equipment, materials and tools are emphasized.

Process-Oriented Flow

The objective of the Process-Oriented Flow program is to convert function-oriented layouts into a series of processes. Layouts are based on the production of product families. Product families here are defined as groups of parts that utilize the same resources (equipment and/or material) in their production. An example of the difference between functional layouts and process-oriented layouts is as follows in *Figure 3-3*.

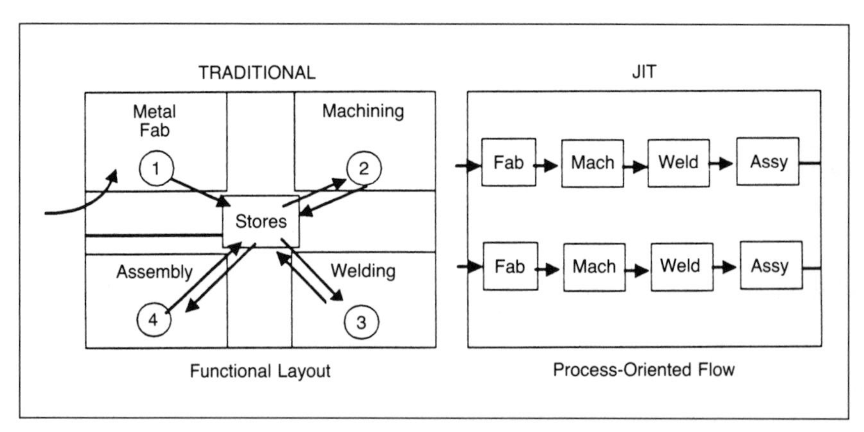

Figure 3-3. An example of the difference between functional layouts and process-oriented layouts.

Process-oriented flows reduce distance, space requirements, material handling, storage facilities, and throughput times. Another important aspect of redesigning the flow of production around processes is the incorporation of cellular manufacturing techniques. Cellular manufacturing typically involves a series of operations organized into one or more process areas called ''cells.'' Cells are often designed in ''U'' or serpentine configurations, in order to better utilize direct labor. By placing workers inside U-shaped cells, the workers can perform more than one operation with relative ease by slowing the production rate. This allows daily production quantities to be varied by changing the number of workers. The variation, in turn, is used to support the matching of production rates to end item demand.

Connectivity is utilized to turn the discrete operations and cells into continuous flows by incorporating pull systems. The pull system is a means of prioritizing and controlling the production process. In the pull system, no production occurs until it is authorized by a pull signal (often referred to as a Kanban) from the subsequent work center. Pull signals can take many forms, such as color-coded containers, cards, golf balls, or even empty squares taped out on the floor. When an empty container or other signal is provided by the next operator, enough product is produced to fill that container. The incorporation of pull signals has often dramatically reduced work in progress (WIP) inventory levels, rework and scrap quantities, and floor space requirements by constructing a ceiling level beyond which no additional WIP inventory is produced.

Advanced Procurement Technology

Technology is commonly defined as a scientific method applied to achieve a practical result. In the area of procurement, the new technology involves understanding that the supplier is an extension of your own company. Your suppliers affect your product costs, customer service levels, quality levels, and ultimate profitability. Because of this situation, it should be the long-range objective of your company to deal with only Just-In-Time suppliers. In this manner, it will be possible to eliminate waste through the entire continuum of your product from raw material to final-product distribution. The development of close, friendly supplier relations is integral to this effort. It will be essential to move from an adversarial supplier relationship to a cooperative one. Contracts must be tied to many aspects of supplier performance other than price. Quality, delivery performance, and other measurements should be included. Data should be shared with suppliers more freely, and on a more timely basis. Electronic Data Interchange (EDI) is often useful in this area.

In addition, contracts should be oriented toward entire commodities whenever possible, rather than individual part numbers. This allows the utilization of fewer suppliers and facilitates long-range contracting for supplier capacity rather than individual part numbers and quantities.

A vendor certification program should be initiated to consistently reduce quality costs. This will require supplier performance evaluation, training, and management commitment.

Finally, suppliers should be drawn into your ongoing improvement processes from design through manufacture, to provide them with insight on how to solve problems and reduce costs.

Improved Design Engineering Methods

Applications of the Just-In-Time philosophy in Improved Design Engineering Methods involve both process and product.

JIT techniques may be applied to the **process** of design engineering to eliminate nonvalue-adding activities in the paperwork flow, and to incorporate manufacturability and quality into the design process. This involves the development and use of "preferred practices" in design, such as minimizing threaded fasteners and labels and standardizing components.

When JIT is applied to the **product** portion of design engineering, programs such as Early Manufacturing Involvement (EMI), Value Analysis (VA), and Current Product Review (CPR) can be utilized to minimize cost while retaining value, quality, and manufacturability in product designs. All of these programs have proven ability to identify lower cost alternatives to product designs while retaining or enhancing product value.

Enhanced Support Functions

Utilization of JIT waste elimination techniques such as the value-added analysis can help to eliminate waste in virtually every support area. In addition, other improvement potentials exist in many of the areas, including:

- *Production Planning*–where the generation and maintenance of production orders, lot sizes, purchase requisitions, and scrap allowances should diminish and may eventually disappear.
- *Production and Inventory Control*–where the expediting of materials, reprioritizing of orders, scheduling of individual work centers, and maintenance of "Hot lists" should become unnecessary.
- *Cost Accounting/Standard Costing*–where traditional cost systems are often converted to process costing systems. Other enhancements may also be required, due to flexible manning levels connected with cellular manufacturing and a lack of WIP inventory to be valued.
- *Production Reporting*–where it becomes possible to report production at fewer and fewer locations in the manufacturing flow. Eventually, bills of material may be backflushed at final assembly, reporting production and relieving inventories all the way back through raw materials simultaneously. Bar coding is often used to further enhance this process.

Employee Involvement

It is difficult to overstate the importance of Employee Involvement in successful JIT implementations. Employee Involvement (EI) is the source of most of the really valuable ideas and suggestions for improvement in every area. It is especially critical in terms of quality, productivity, and design. It can be implemented in parallel with other JIT activities, and it usually yields significant immediate returns. EI facilities "buy-in" to the program and unifies improvement efforts in a single management structure. EI is the epitome of participative management, and the heart of program synergy.

Employee involvement requires that problem-solving work groups be established, along with a steering committee to guide their efforts. Work groups are trained in effective problem-solving techniques, cost/benefit analysis, and working effectively as a group. The steering committee regularly reviews the progress of the groups as they continuously identify, prioritize, select, and resolve problems. Once initial JIT implementation activities are completed, the program focuses on EI to constantly apply the JIT philosophy of waste elimination in the work environment.

Now that we have completed an overview of the elements of JIT and the implementation methodology, we will take a more detailed look at the individual components comprising the JIT "emerald."

QUALITY

4

Defining and promoting quality has become a tremendous growth industry. It has been most widely described as "fitness for use," with related phraseology about uniformity and dependability. Perhaps quality is most appropriately defined in terms of consistently satisfying customer expectations, while minimizing cost. Of course, customer satisfaction implies "fitness for use." However, it involves more. The best quality programs proactively identify and address both existing and potential quality problems. Generally speaking, quality can only be achieved in an environment where it is a recognized and accepted part of each individual's responsibility. Recognition and acceptance, in turn, are generated by strong top management commitment and comprehensive awareness/education efforts.

Achieving Commitment to Quality

Achieving commitment to total quality control (TQC) is neither an easy nor finite task. Like the other aspects of Just-In-Time, it requires ongoing effort.

The initial "sale" of TQC should revolve around tangible benefits. These benefits can be utilized to construct sound performance measures for later TQC activities. However, their primary use at early stages is to stimulate interest and convince management from a "dollars-and-sense" standpoint that the program is a good idea. Pilot programs may be initiated to "prove" TQC viability, but this takes time and momentum may be lost. Other proofs may be offered, such as published benefits from similar programs and formal opportunity assessments by professional quality consultants.

When top management has agreed to provide at least tentative support, they must then introduce the quality control program to key personnel and give the charter to the TQC team. At this point, a complete analysis must be done on current quality costs and improvement potential. As the program continues, progress should be monitored on a regular basis, with successes reported to the entire organization to promote the TQC efforts.

Customer-Defined Quality

Quality, as stated in an earlier chapter, must be defined as it relates to the customer. Generally, there are two types of customers.

Those generally regarded as customers are the ultimate recipients of the product or service offered by the organization. Determining what these customers perceive to be the characteristics of value is usually accomplished through surveys and interviews. It is essential that those aspects of the product deemed to be critical by the customer are quantified and developed into a formal set of guidelines for monitoring and control.

The second type of customer is the internal customer, the subsequent user of the product or service within the producing organization. This includes downstream manufacturing operators and support personnel, such as analysts who use production data generated by direct labor employees. One of the most effective ways to monitor and control quality levels related to internal use is to implement procedures called "point-of-use inspection," "next operation inspection," and/or "quality at the source." The point of these programs is to train operators to inspect critical quality elements of their own, as well as those immediately preceding their individual operations. Typical quality issues at downstream operations include problems with fit, due to out-of-tolerance processes. Experienced workers usually have no trouble identifying quality problems created upstream, and they will do so readily in interviews. Frequency studies will be helpful in prioritizing the problems, after which measurement criteria must be developed to consistently monitor their occurrence.

The Two Major Aspects of Quality Control

When quality has been defined in such a way that its discrete aspects are measurable and clearly communicated, quality control boils down to two ongoing processes: 1) Identification and resolution of defects and, 2) Defect prevention.

Identification and resolution of defects generally follows a sequence of events like this:

1. *Statistical detection*–where critical specification inspection is utilized to identify defective parts, and defect levels are chartered to document trends.
2. *Problem investigation*–where a cursory attempt is made to identify the root cause of the defect. If the root cause is obvious, it is remedied at this point and the resolution process ends.
3. *Problem solving*–which involves brainstorming and cause-and-effect analysis–is applied to the defects when root causes cannot be identified by cursory investigation. At this point, the underlying cause is determined to be systemic or nonsystemic. If it is nonsystemic, it is resolved and the process ends.
4. *System Identification*–If the problem is systemic, the system responsible must be identified. Data gathering is performed to support the system identification and an analysis of how the problem occurred. The organization responsible for the causal system is brought into the resolution process, and the problem is resolved.
5. *Verification*–which involves confirmation activities to insure that the corrective action adequately resolved the problem in question.

Defect prevention in the manufacturing environment generally involves the following sequence of events:

1. *Identification of critical processes*–which is basically the identification of those processes responsible for achieving critical specifications (specs). For example, if the "customer" who is trying to paint the material sent to him cannot do so when it is burred, then the process by which the material is machined must be monitored to prevent burrs from occurring. A burr-free surface is the critical spec, and machining is the critical process in this case.

2. *Identification of critical process elements*–which is a methodical analysis of the elements of each critical process that is used to determine what aspects of the process may be the cause of the problem. In the burr example, the critical element might be alignment of the boring tool.
3. *Ensuring fail-safe critical elements*–which involves the development of ways to prevent defect-creating conditions in the critical elements. Returning to the burr example, this might mean building a fixture that will not allow variability in the alignment of the boring tool.
4. *Statistical process control*–where a fail-safe method is not appropriate, critical process elements must be monitored and plotted graphically over time. This statistical control can then be used both to ensure that process capabilities that are deteriorating are addressed before they create an "out-of-spec" condition and to continuously narrow capability variation.

Tools of Quality Control

There are many tools available to assist in the quality control effort, including frequency charts, scatter graphs, Pareto charts, process control charts, "$\overline{X}$" and "R" charts, fishbone diagrams, check sheets, histograms, relational diagrams, affinity diagrams, and matrix diagrams. A few of the more effective ones are considered here.

Frequency charts, which are among the simplest tools of quality control, monitor defect frequency by part. Each time a defect is identified, it is written on a blackboard near the work center. At the end of each day, all defects are recorded in a logbook. On a monthly basis, the defects are sorted and graphed by part number for frequency.

Pareto charts are not only easy to construct (especially with spread-sheet capability), they are also extremely flexible in their application. In the case of the frequency data described in the previous section, for example, over time a Pareto chart will help to quickly identify the 20% of parts which contain 80% of the defects. A Pareto chart can be constructed by first listing defect data as shown in *Figure 4-1*, and then sorting it into descending order as shown in *Figure 4-2*.

The top 20% of the part numbers in this case consist of part numbers 1659 and 2100. These two part numbers also represent 80% of the defects. The remaining 80% of the part numbers represent 20% of the defects.

Pareto's Law, then, as demonstrated here, simply implies that 20% of any given population of occurrences represent 80% of the total value of those occurrences. Although Vilfredo Pareto (an Italian-Swiss engineer and ecomomist who lived from 1848 to 1923) introduced this law in the context of income distribution, it has been widely applied in a broad range of settings since its inception.

One of the most popular uses of Pareto's Law is the valuation and cycle counting of inventories.

Process control charts are utilized in a number of ways in the manufacturing environment, including:

- Plotting nonconformances in final assembly.
- Plotting critical spec ranges in part production.

Part Number	Frequency
C1239	1
A1659	64
C1113	1
C1294	9
C1299	1
A2100	16
C2155	1
C1667	5
C2305	1
C3121	1

Figure 4-1. A Pareto chart.

Part Number	Frequency
A1659	64
A2100	16
A1294	9
A1667	5
C1239	1
C1113	1
C1299	1
C2155	1
C2305	1
C3121	1

Figure 4-2. A Pareto chart in descending order.

- Plotting critical spec ranges of incoming material.
- Plotting tool wear.

For the purpose of this text, consider an example of how process control charts are used to plot critical specs on parts during production, using an "X" or an "R" chart.

$\overline{X}$ and R charts plot averages and ranges of critical specs during production. $\overline{X}$ is the graphic representation of occurrence averages, and it is used to smooth out the effects of occasional variations to display a picture of general trends. R is a plot of actual occurrences, which displays the range of their travel. Both X and R are typically charted on the same sheet, with lot numbers and sample spec data recorded for each occurrence. The plots are graphed relative to upper control limits (UCL) and lower control limits (LCL) set within the designated spec limits. UCL and LCL are usually determined by the standard deviations from a centerline or true spec.

As measured occurrences begin to approach either control limit, corrective action is considered to bring the process back under control.

In a sound TQC environment, control limits are constantly being narrowed and achieved. (For a sample control chart, see *Figure 4-3*.)

The "fishbone" or "cause-and-effect" diagram, originally called the Ishikawa diagram, is a very popular technique for cause-and-effect analysis in problem solving.

This technique seems to be one of the best available methods for structuring brainstorming about the causes of any existing problem. It is a kind of graphic "outline" of potential causes and causal relationships.

The problem is listed at the far right end of the "fishbone," with at least four spurs from the main trunk. In this text these four spurs are usually referred to as "materials," "methods," "manpower," and "machinery"—just to get things started. Almost any underlying potential problem causes can be listed in this fashion.

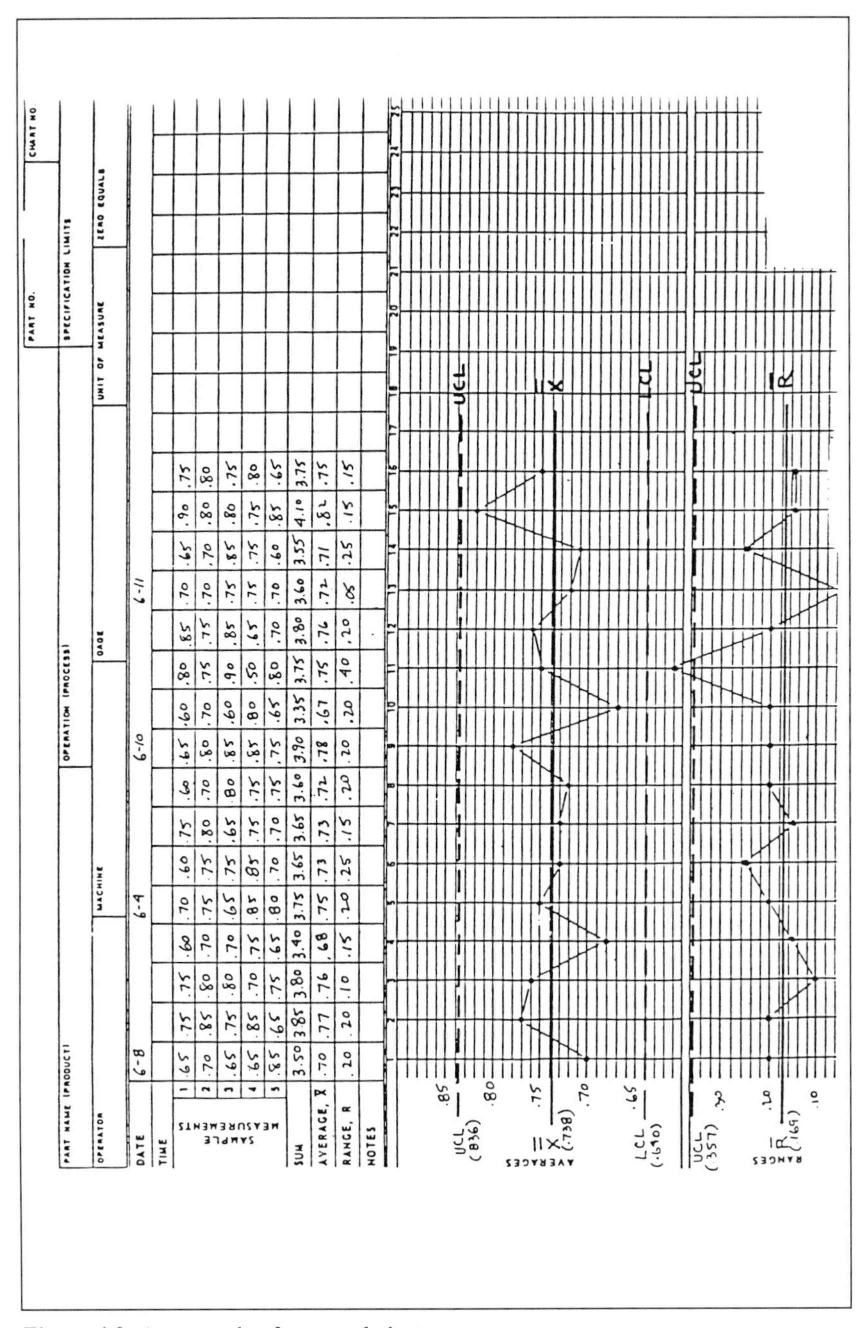

DATE	6-8				6-9				6-10				6-11			
TIME																
SAMPLE MEASUREMENTS 1	.65	.75	.75	.60	.70	.60	.75	.60	.65	.60	.80	.85	.70	.65	.90	.75
2	.70	.85	.80	.70	.75	.75	.80	.70	.80	.70	.75	.75	.70	.70	.80	.80
3	.65	.75	.80	.70	.65	.75	.65	.80	.85	.60	.90	.85	.75	.85	.80	.75
4	.65	.85	.70	.75	.85	.85	.75	.75	.85	.80	.50	.65	.75	.75	.75	.80
5	.85	.65	.75	.65	.80	.70	.70	.75	.75	.65	.80	.70	.70	.60	.85	.65
SUM	3.50	3.85	3.80	3.40	3.75	3.65	3.65	3.60	3.90	3.35	3.75	3.80	3.60	3.55	4.10	3.75
AVERAGE, $\bar{X}$	.70	.77	.76	.68	.75	.73	.73	.72	.78	.67	.75	.76	.72	.71	.82	.75
RANGE, R	.20	.20	.10	.15	.20	.25	.15	.20	.20	.20	.40	.20	.05	.25	.15	.15
NOTES																

Figure 4-3. An example of a control chart.

Consider the example shown in *Figure 4-4*:

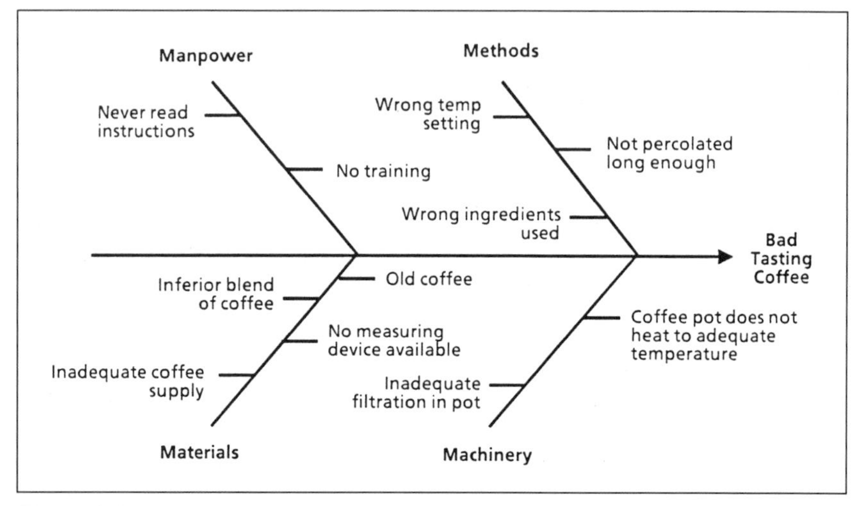

Figure 4-4. An example of a Fishbone diagram.

Quality Control Responsibilities

Quality is everyone's responsibility. To think otherwise is to ignore recent historical lessons provided by international competition, and to insure the demise of manufacturing in the United States.

In order to manage identification and resolution of existing defects, and defect prevention activities, a quality organization must be constructed with these primary participants:

Top management–who must allocate adequate resources (money, manpower, their own time, etc.) to support ongoing and future needs of the TQC program. They must present an attitude that quality is a top priority to all of their employees, customers, and suppliers. The development and maintenance of challenging and appropriate quality standards must be supported by top management and then reinforced by regular review meetings and management attention to quality levels and performance of goals.

The quality control manager–who must structure, initiate, and maintain the TQC program. Included in the QC manager's responsibilities are the construction of quality objectives for each appropriate organization level or unit; provision of adequate manpower, equipment, and training to support the organization in achieving its objectives; and monitoring and reporting the results of these activities to top management. The QC manager will also need to move the organization from a "defect resolution" orientation toward one of "defect prevention" through the operation of the TQC program. Finally, the QC manager will need to oversee the activities of any ongoing quality circle program.

Quality control engineers–whose responsibilities include the development of any required procedures for operator-controlled testing and/or inspection, and inspector/operator training. The QC engineer is also often responsible for designing and specifying test/inspection equipment and for overseeing the collection of defect data. QC engineers should assist both operators, in identifying underlying defect causes, and inspectors, in testing for process reliability. Finally, QC engineers should oversee—or at least actively participate with—purchasing and engineering departments in the review of supplier capability.

Other department managers–need to play a very aggressive role, if not the key role, in establishing their individual department's quality goals. They must "buy in" to these goals and communicate them with a sense of urgency to their subordinates. Then these department managers must monitor their own department's performance of these goals, just as any other critical spec would be monitored, and initiate corrective action when goals are not being met.

Quality circles–are groups of employees (generally 5 to 10 employees per circle) who regularly meet to address quality problems in their work areas. Usually, these employees are organized by process or geographic proximity. They track occurrences or defects in their areas (see Pareto diagrams, earlier in this text) and then address them as a group. Their goals are to identify, investigate, and eliminate defects in their areas.

Specifically, the responsibilities of quality circle members are to track defects (expressed typically as deviations from any critical spec) and to participate on a regular basis in the investigation and resolution activities associated with these defects. In addition, quality circles frequently have to report on their activities and present improvement proposals to management for approval. Finally, quality circle members are often charged with implementing corrective actions, monitoring results, and reporting these results to management once they have been verified.

While quality circles are a form of employee involvement, they are distinct from the EI organization that will be discussed later. Quality circles usually concentrate solely on quality-related issues. They are more permanent groups, which retain the same members as more and more problems are tackled. Quality circles became most popular during the early 1980s. Since then, they have made lasting and valuable contributions to many companies where management commitment has been adequate and ongoing.

The Quality Development Process

The quality planning or quality development process of incorporating all of the quality techniques, responsibilities, and activities into a TQC program that will support Just-In-Time activities must be outlined.

JIT implementations pose special challenges related to quality, since fewer parts are available to use as buffers when defects occur. As a result, quality-related activities are among the first activities to be undertaken in a sound JIT program, and they are among the most critical to overall success.

The overall TQC program can be visualized as three major elements: 1) Quality Planning, 2) Quality Control, and 3) Quality Improvement.

The objective of quality planning is to create a process that will attain appropriate, customer-defined quality goals under all projected operating conditions. In this process, internal and external customers are identified. Customers are then interviewed to determine desired critical quality characteristics. These characteristics are broken down into specific features, from which critical specifications are derived for both existing and new products. (The use of ''products'' here means all products and services produced by the organization.) Processes are then identified which will be required to produce the desired specifications at the required quality levels on a consistent basis, while at the same time minimizing cost. Costs of quality are then defined and benchmarked to provide baseline data for future improvement activities. (For typical costs of quality, see *Figure 4-5*.) Process capabilities are verified, and all processes are documented.

The objective of quality control is the continuous achievement of quality goals during manufacturing and support operations. In order to accomplish this, critical specs are communicated, along with their associated processes and measurment methods/devices. Performance standards are established, and training is carried out. Performance is measured on an ongoing basis and reported. As processes and/or specification control limits are reached, corrective action is initiated to compensate for these situations.

The objective of quality improvement is to continually increase defined levels of quality over previous performance. Typical activities undertaken in this area include variation research and designed experiments to determine root causes for variation. Generally, extremely critical specs are addressed first, since this process can be expensive and time-consuming.

- Internal Failure Costs
  - Scrap
  - Retest
  - Yield Loss
  - Rework
  - Downtime
  - Disposition
- External Failure Costs
  - Complaint Adjustment
  - Returned Material
  - Warranty Charges
  - Allowances or Discounts
- Appraisal Costs
  - Incoming Material Inspection
  - Inspection and Test
  - Test Equipment Maintenance
  - Resource Consumption (Destructive testing, electricity, etc.)
- Prevention Costs
  - Planning Activity
  - New Product Review
  - Training
  - Process Control
  - Quality Data Collection/Analysis
  - Quality Reporting
  - Preventive Maintenance

Figure 4-5. Typical costs of quality.

SIMPLIFIED SYNCHRONOUS PRODUCTION 5

As mentioned earlier, the objective of SSP is to match production rates as closely as possible to the actual rate of sales. Ideally, this will be on a sales-per-day basis. However, many organizations start out at a monthly sales level and slowly become more and more exact.

SSP can only be accomplished as lot sizes are reduced to minimal levels, with frequent changeovers to meet final production schedules. The existing ''batch'' production approach in traditional manufacturing settings precludes SSP by its very nature. It takes longer to produce larger batches before the next operation is performed. By the time this extended production time is applied to each operation in the production routing, the product's throughput time is entirely too long to allow sales to be matched closely. Other disadvantages to batch production include the excessive work in process inventory levels generated between operations, and increased levels of scrap and rework. (If there are 500 pieces of WIP inventory in batches between a faulty work center and the point at which the defect is detected, there are 500 pieces to be reworked or scrapped. If, however, there are only 10 pieces of inventory spread across the same number of operations, the rework efforts or scrap levels are proportionately reduced.) In order to change between products more frequently without penalizing output levels, setup/changeover times must be reduced.

Setup Time Reduction

Setup/changeover time is defined as all of the elapsed time from production of the last good piece on the old setup until the production of the first good piece on the new setup. This includes all time waiting for materials, tools, setup people, and first-piece inspection.

The first principle of setup reduction is to put all tools, dies, fixtures, gages, and other material required for the setup in an assigned location near the machine. Operators should not lose time searching for these materials, and the best way to prevent this is the old housekeeping technique of having ''a place for everything and everything in its place.'' Tool outlines painted on peg board, numbered shelf locations, etc. are all acceptable methods for locating and organizing setup materials.

Working in conjunction with this principle is the concept of utilizing as few of these devices as possible. Some organizations actually measure the performance of their engineering staff by how few tools are required for setup and changeover.

The next principle involved in reducing time expended in setups/changeovers is often referred to as internal/external. It is simply the practice of doing every setup element possible while the previous job is still running. This minimizes activities external to actual production by internalizing elements like retrieving tools, preheating molds, and positioning containers. As much as 50% of setup times can be eliminated using this technique.

The application of these first principles may necessitate the purchase of some duplicate tooling, gages, etc. to eliminate time spent searching for and retrieving them. The cost involved is typically recovered very quickly and should not be a serious deterrent to setup time reduction activities.

The next principle of setup reduction is the use of quick-change couplings, clamps, and other fast-change devices in place of threaded fasteners. These units have increased markedly in availability and use in recent years. Racheting fasteners, locator pins, and fixed-position stops can dramatically reduce setup times while at the same time improving product quality by reducing variation.

The last principle involved in setup reduction is the elimination of setups altogether by dedicating equipment to the production of a single product. Here again, there are limitations in the area of flexibility and additional cost. However, setup costs are virtually eliminated in this process.

Uniform Work Loading

SSP requires more than small batch sizes. It also requires uniform work loading through the production routing.

Uniform work loading means matching the cycle times for individual operations within the production routing, so that production occurs as though to a drum beat. With each beat of the imaginary drum, material moves simultaneously between work centers. In this manner, WIP inventory levels are minimized and throughput is improved. To accomplish uniform work loading, multiple machines may be required for a single operation. Another technique is to balance the work load by utilizing individual direct-labor employees to operate more than one machine. For an example of uniform work loading, see *Figure 5-1*. In this Figure, production moves downward through the routed operations at a uniform rate, even though different levels of equipment and manpower are utilized from operation to operation to achieve matched cycle times.

Matching cycle times along the production routing becomes more complex as multiple jobs run across the same work center. The first steps in analyzing the production activities for a given work center in this situation are:

1. The calculation of cycle time (or "drum beat") to determine the number of seconds per unit produced for the total number of parts required from the work center in a designated period. Consider the example in *Figure 5-1*.

   Production spanning 20 days per month, at 8 hours per day results in requirements for 1450 per day or 182 per hour. Therefore, to achieve the monthly demand of 29,000 pieces indicated in *Figure 5-1*, the cycle time at this work center would be 20 seconds per unit (3600 seconds per hour divided by 182 pieces per hour).
2. When the cycle time has been calculated, the work load must either be engineered in such a manner that all of the parts on this work center may be produced (including setup and changeover) in this 20-second increment, or it must be redistributed to achieve this cycle time by off-routing.

An additional level of complexity is incurred as multiple end item usages are involved for the components manufactured on the work center. Consider the example in *Figure 5-2*.

Monthly Production Requirement	Part Number
15,000	A1659
10,000	A2100
2,000	C1294
2,000	C1667
29,000	

Figure 5-1. An example of uniform work loading.

Part Number	Monthly Production Requirement	Quantity Required per Hour	Required Unique Part Cycle Time	Required Cycle Time for Parts Common to A1659 & A2100	Required Cycle Time for Parts Common to A1659 & C1294
A1659	15,000	94	39 sec.	23.04 sec.	
A2100	10,000	63	57 sec.		
C1294	2,000	13	277 sec.		34 sec.
C1667	2,000	13	277 sec.		
	29,000	183			

Figure 5-2. An example of increased complexity in uniform work loading.

The point of this *Figure* is to demonstrate the increased complexity that occurs in uniform work loading as work center production cycle times are calculated based upon multiple products and multiple components per product. The best way to deal with this situation is to equalize the cycle time across all of the components produced on any given work center.

Housekeeping

Housekeeping is an often underestimated aspect of SSP. Poor housekeeping can result in lost tools and equipment, lost time, poor safety, poor quality, and poor morale. It is the most apparent evidence of how management feels about its employees, and how employees feel about the quality of their product. It can be an important first step in initiating employee involvement and in demonstrating that management cares about its employees. Providing simple cleaning tools, relocating equipment, and improved lighting over work centers are examples of the kind of changes management could make to enlist employee support.

The first principle of housekeeping, as mentioned earlier, is "a place for everything, and everything in its place." This adage is valuable in terms of more than tooling. Materials for production and/or testing, scrap and offal, dunnage, packaging materials, and containers are all included. The mental effort required to plan for and allocate floor space or racking to these items goes a long way toward helping management to properly assess what is and is not needed at each work center.

A corollary to this principle is visibility. Everything at the work center should be properly identified and placed where it can be easily seen. There should be no last-minute surprises involving missing or broken tools, gages, fixtures, or materials during production. Keeping all of these items out where they can be readily seen and identified is the best way to avoid these unpleasant occurrences.

The next principle of housekeeping is cleanliness. Operators need to keep their work areas and equipment as clean and free of debris as possible. Dirt, dust, excessive oil, and other foreign materials can cause equipment to malfunction and can negatively impact employee morale.

Finally, as operators are performing regular machine "wipe-downs," they should check for oil leaks, squeaks, and other signs of potential machine failure. Such incidents should be addressed as part of an ongoing preventive maintenance program.

Preventive Maintenance

Preventive maintenance (PM) is another important aspect of simplified, synchronous production. It is a program of systematic inspection, detection, and prevention of failure in production and support equipment that reduces delays, supports employee safety efforts, and ultimately reduces operating costs. Preventive maintenance can be perceived as falling into two categories:

- *Schedule or planned maintenance*—which occurs periodically according to elapsed time, hours of equipment operation, or equipment age.
- *Monitored or variation-oriented maintenance*—which occurs as it is triggered by variation increases in the process (detected via SPC) or observation of potential problems (oil leaks, etc.).

A recent article by Joe Gufreda and Paul Seus of Ernst & Whinney outlines these indications of the appropriateness of preventive maintenance:

- PM is required for safety, or to satisfy insurance regulations.
- PM is less costly than breakdown.
- Breakdown would further damage the equipment.
- Breakdown would result in critical customer service problems.

Preventive maintenance programs may be implemented in one of two ways:

- Company-wide program.
- Pilot program and remaining factory conversion.

The pilot program, which is recommended, may be originated by department, equipment type, or product line. Initial program successes can then be used to sell the concept, and provide available training ground.

Culture Changes Associated with SSP

Simplified, synchronous production is an integral part of Just-In-Time operations. Without simplifying and synchronizing individual production operations, it will be impossible to eliminate waste in the manufacturing process. The process of converting from a batch orientation with protective levels of WIP inventory "buffers" between operations to a very tight, lean production operation that is extremely vulnerable to shortages is very difficult—and often very painful. It is important to note that this process can be done incrementally and "staged" into place, beginning at the end of the assembly line and working backward a little at a time toward the procurement of raw materials. Another approach, which is preferred, is the "pilot line" approach. This technique is most easily visualized in conjunction with rearranging the shop floor, as described in the next chapter of this text under "Process-Oriented Flow." Utilizing the pilot line approach, the entire process is synchronized in terms of cycle times on one product line, in an environment where a fitful start-up will not jeopardize remaining production operations. The major problems are identified and resolved in this less-threatening environment, and an initial success is generated. The knowledge thus gained from the pilot line is then more easily transferred to remaining lines, until the entire program is converted to synchronized production.

PROCESS-ORIENTED FLOW

6

As described previously, the objective of orienting product flow to the production process is to convert the manufacturing layout from traditional functional grouping to a more straight-line process or series of processes.

Ideally, layouts of this kind are developed around a product family. Product families are groups of parts that share the same resources to produce them, and /or material from which they are fabricated. Part families are different in each environment. For example, an assembly operation might well regard all of the metal stampings that it purchases as a single commodity or part family. However, the stamping house breaks down the stampings into backing plates, brake parts, and spring retainers because these divisions require different machines to produce them. When involved in the identification of product families, the first question is usually something like "What parts run across the same series of operations?" Oftentimes, true to Pareto's principle, relatively few part families comprise the bulk of production volumes. This approach frequently supports the division of what are classically referred to as "job shops" into two smaller business units. One is centered around the more repetitive (high-volume) product lines, and the other is made up of very low-volume, make-to-order products.

A "focus factory" approach, a kind of "factory within a factory," allows business units to be formed around one or a few product lines in order to concentrate resources in an efficient way on waste elimination and productivity. It provides some degree of autonomy to this business unit, especially in terms of its own operating budget and P&L responsibility. When focus factory operations are established, it becomes much easier to realign product flow, since most support resources are oriented toward a semi-autonomous operation anyway.

When product families are identified, and Paretoized to distinguish the high-volume lines, the operations and equipment required to produce each product line is documented. Usually, it is best to develop a rudimentary flow chart for each product line and remove all nonvalue-adding activity from the process in the course of flow chart development. Nonvalue-adding activities can be defined as any activity that is involved in the production process but that does not add value to the product. Implementers typically identifies nonvalue-adding activities by literally following the part from operation to operation as it is produced and recording each time the product is handled. Travel distances and inventory levels are also recorded, as shown in *Figure 6-1.*

In this case, we can see that only 2 of 13 material handling occurrences add value to the product.

Often more than one part is produced in a similar manner, across many of the same work centers, and in a very similar sequence. When this happens it is helpful to combine the flow charts for all of the parts involved, attempting to maintain the cleanest possible flow by minimizing the number of times flow lines on the charts cross each other.

Activity	Value Added	Description	Travel Distance	Inventory
A	No	Receive, inspect, label	---	100 pcs
B	No	Move to storage area	480′	---
C	No	Store	---	1,000 pcs
D	No	Move to blank machine	240′	---
E	Yes	Blank	---	250 pcs
F	No	Move to queue at curl machine	90′	---
G	No	Queue	---	200 pcs
H	No	Move to curl machine	15′	---
I	Yes	Curl	---	50 pcs
J	No	Move to shipping area	250′	---
K	No	Stage	---	1,500 pcs
L	No	Package parts	10′	100 pcs
M	No	Ship	25′	---
Total	2/13		1,110′	3,200 pcs

Figure 6-1. An example of recorded travel distances and inventory levels.

When these flows are grouped logically together in such a manner as to minimize the number of flow charts required, as well as the number of crossed flows, a good start has been achieved toward a model of future physical operations.

At this point, the concepts of cellular manufacturing should be applied to the model. Cellular manufacturing involves the organization of production operations into "cells" that are in close physical proximity. Cells are usually U-shaped or serpentine, to allow operators to move back and forth between work centers easily, so that more than one operation may be performed by a single operator. (For an example of U-shaped and serpentine cells, see *Figure 6-2*.)

In cellular manufacturing environments, the product never leaves the cell between operations. All WIP is between the work centers, and it is minimal. Another tremendous advantage of cells is the ability to vary manning levels within the cells in order to adjust the rate of production. Cells normally manned to produce 100 pieces per hour utilizing 6 operators, for example, can be manned with 3 people to produce 50 pieces per hour.

Between cells and other noncellular operations, WIP inventories can be minimized by using "pull" production systems. Pull systems differ from traditional push systems in this fashion:

Push systems launch production orders for components and assemblies based on system due dates and estimated manufacturing throughput times, adjusted by lot sizing logic and scrap allowances. As a result, large lot sizes and frequent shop production priority changes occur. Coordination is lost, and component parts are not completed at the same time to feed assembly operations. WIP

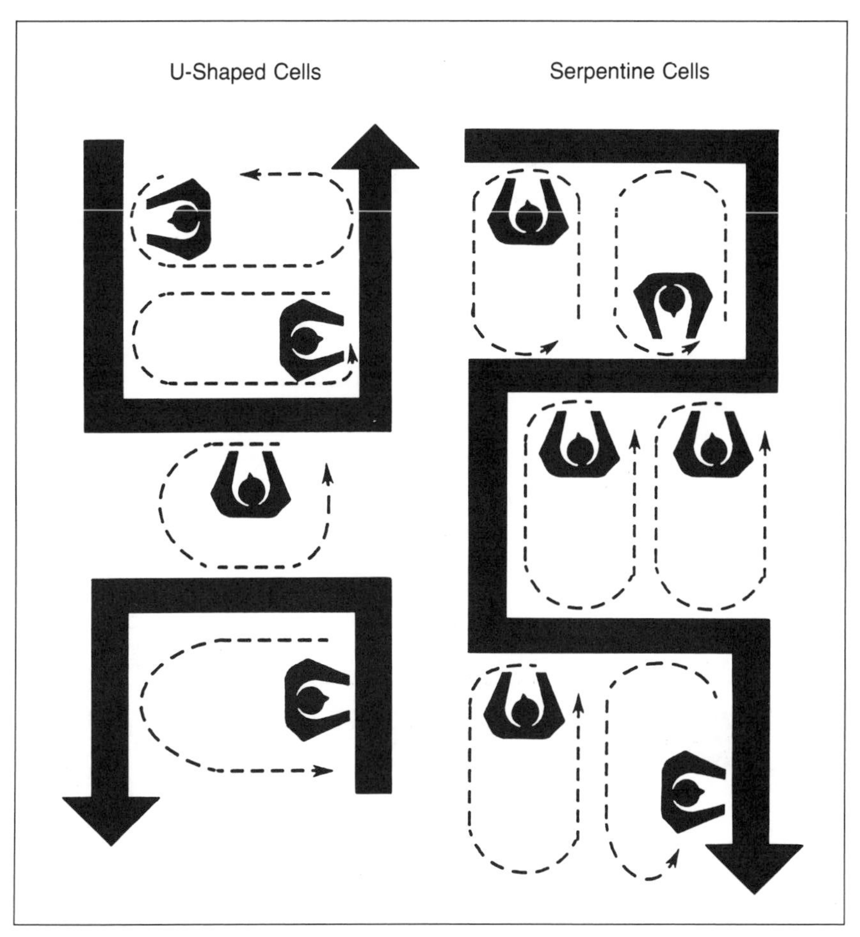

Figure 6-2. U-Shaped and Serpentine Cells.

inventory builds, and risk increases for large quantities of defective material between operations prior to detection. Consider the example shown in *Figure 6-3*:

In this example WIP inventory is at 2,500 pieces, and defective material produced at work center number 1 would grow to roughly 1,000 pieces before it would be detected by the operator of machine 2.

Pull systems establish WIP inventory "ceilings" between work centers, allowing no additional production when "ceiling" levels are reached. The production sequence and volume are controlled by pull signals or "kanbans." Types of pull signals include empty containers, cards, color-coded golf balls, and empty squares taped out on tables or floors. Connectivity is the key here; it requires modification to the last example, as shown in *Figure 6-4*:

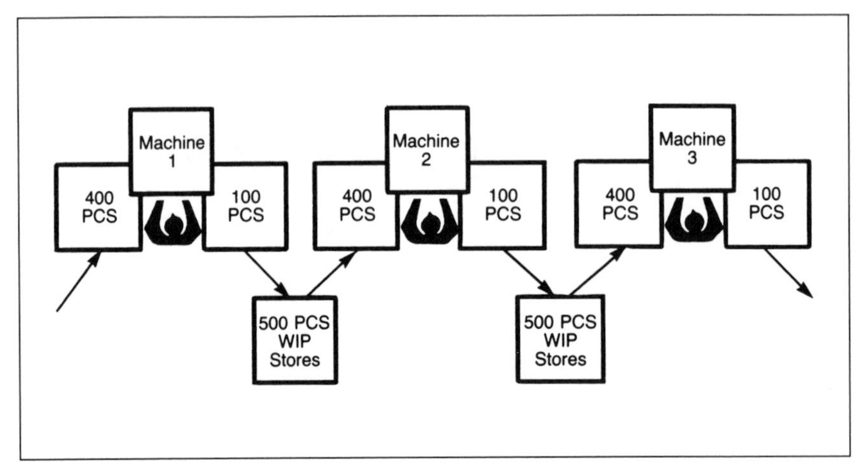

Figure 6-3. An example of a Push System.

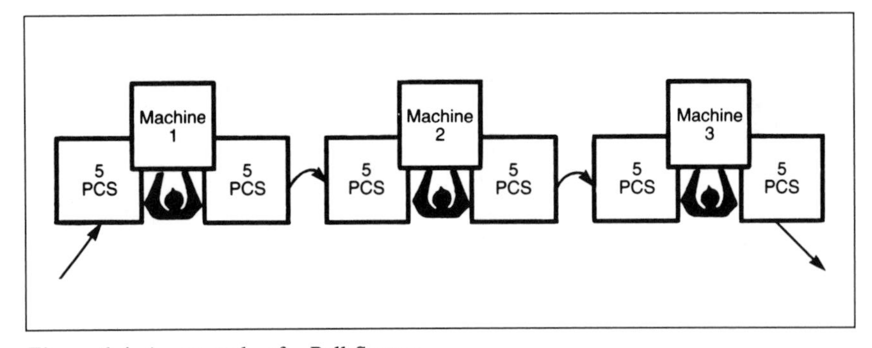

Figure 6-4. An example of a Pull System.

In this environment, WIP inventory is reduced to 30 pieces. In addition, no more than 10 defective pieces will be produced on machine 1 before the operator of machine 2 detects the problem.

Pull system production priority sequencing can best be visualized utilizing the following model, as shown in *Figure 6-5*:

As production occurs in this setting, a defect or other interruption in the production of subassembly B would result in one of two scenarios:

- *Push System*—Production continues unimpeded on C, D, E, F, and G building WIP inventories. In addition, since a shortage of subassembly B will prevent production of final assembly A, production priorities on all other A subassemblies and components will need to be de-expedited; other production priorities will need to be established for these work centers, with related expediting on these parts.

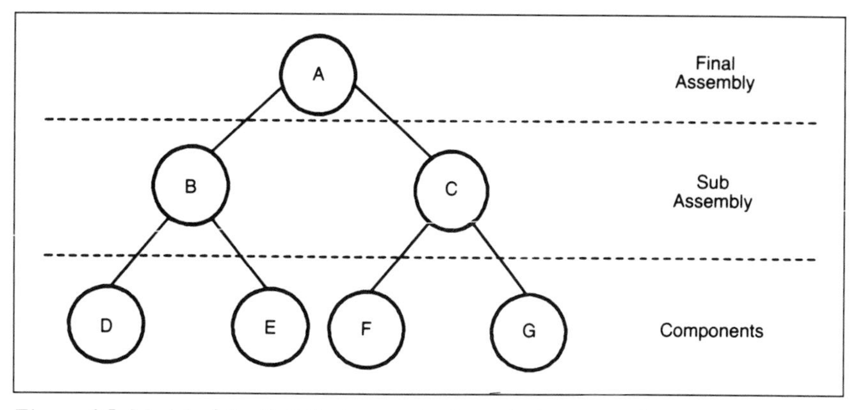

Figure 6-5. Model of the Pull System production priority sequencing.

- *Pull System*—Production stops on all subassemblies and components since kanbans (or pull signals) are no longer being issued from B or A. Everyone from these work centers converges on work center B to identify and resolve the problem, and production is resumed.

With a model constructed of streamlined physical operations and the incorporation of pull systems, production equipment can be realigned to achieve substantial productivity improvements. The movement and reorganization of production equipment into connected and/or cellular manufacturing is another instance where it is often advantageous to utilize the pilot line concept. A pilot program involving the entire manufacturing process of a given product line can be set up in order to prove the viability of these techniques, before leveraging them into remaining factory operations. This allows the ''bugs'' to be worked out of the process before a large percentage of production is risked, and it provides an excellent training opportunity for operators and support personnel.

ADVANCED PROCUREMENT TECHNOLOGY (APT) 7

As mentioned earlier, it is important when applying JIT philosophy to the purchasing operation to regard suppliers as an extension of your own company. Supplier operations, like your own, are merely part of a continuum stretching between raw materials (metal ore, petroleum, etc.) and the ultimate consumer of the product. The objective, then, of APT is to integrate manufacturer and supplier operations. There are a number of aspects involved in this concept, including:

1. Data Sharing
2. Improved Ordering Devices and Methods
3. Commodity Contracting
4. Comprehensive Supplier Performance Measures
5. Supplier Certification
6. Buyer Evaluation Criteria
7. Joint Cost Reduction
8. Supplier JIT Training
9. Revised Purchasing Organization

Each of these aspects will be examined in more detail.

Data Sharing

Sharing data with suppliers is one of the best ways to foster trust. In addition, it allows suppliers to be more responsive by providing them with the earliest possible warning regarding demand changes, design changes, and production schedule changes. Sharing critical data with suppliers can only be accomplished safely when the supplier base is minimized, so that there is little risk of competitors getting confidential data.

Experience suggests that most manufacturing organizations are still uncomfortable with fewer than two sources on most commodities. "Acts of God," such as fires, floods, etc., are one of two reasons usually cited; the other is a potential labor dispute at the supplier. It is usually recommended that if two suppliers are maintained on a commodity, the volume of purchases should be split at roughly 90% and 10%. This allows greater leverage at the major supplier and still retains the second supplier safety valve.

The data that can be shared with suppliers to improve relations and responsiveness includes:

- Production schedules–so that suppliers have the longest possible lead time to meet schedule fluctuations.
- Sales forecasts–so that suppliers may better forecast their own sales and related purchases of raw materials.
- Design information–so that suppliers can more quickly respond with raw material procurement changes, process changes, and feedback about resulting costs.

Data sharing can be improved by delivering it in a more timely manner. One of the most effective recent improvements in this regard is the advent of Electronic Data Interchange (EDI). EDI application will be discussed in more detail in the subsequent chapter on enhanced support functions, but for now it is merely important to recognize that EDI can dramatically accelerate the flow of important data between manufacturers and their suppliers. EDI accomplishes this by electronically linking the suppliers' computers with those of the manufacturer.

Improved Ordering Devices and Methods

Electronic data interchange, as described above, can certainly improve data flow times and accuracy after the data is identified and defined. However, there is an additional element of the ordering of purchased parts that must be considered, namely the pull signal.

The use of kanbans or pull signals in the manufacturing process was described in an earlier discussion of process-oriented flow. However, when considering the supplier to be an extension of your own manufacturing operation, the pull signals may be transmitted electronically via telephone or EDI, or they may be something as simple as empty containers returned to the supplier. As batch or lot sizes are reduced, this implies more frequent delivery and supplier responsiveness.

In order to support Just-In-Time delivery on the part of suppliers, the ordering of purchased parts and materials can be done in such a way that contracts span one- to five-year periods, with gross requirements over the life of the contract stated up front and individual releases (or pull signals) consuming contracted requirements on an ongoing basis.

Commodity Contracting

In addition to longer contract time spans, it is often advantageous to orient contracts to supplier capacity for the production of an entire commodity.

As recalled from earlier discussions, a commodity is defined for the purposes of this text as a group of parts that require the same raw materials and/or production resources to produce them. (One commodity that has proven to be an excellent starting place for many manufacturers' purchasing improvement efforts is threaded fasteners.) When individual part numbers are each covered by a separate contract, relatively small part volumes and fluctuating demand levels result in poor outcome in terms of both cost reduction and cooperative efforts in the areas of delivery performance and quality.

By combining the requirements for all of the parts within a given commodity (threaded fasteners, sheet steel, wiring harnesses, etc.), it is often possible to contract for a percentage of supplier volume, thereby obtaining these advantages:

- Leverage–with the supplier to gain price, delivery, and quality performance improvements.
- Flexibility–to change the requirements for specific parts numbers much "closer in," to adjust for changes in demand.
- Efficiency gains–through the generation, maintenance, and monitoring of fewer contracts.

The combination, then, of longer term contracts with fewer suppliers for entire commodities is a very desirable mix for purchasing departments seeking to gain significant long-term cost reductions and efficiencies.

Comprehensive Supplier Performance Measures

There are at least two fundamental reasons to measure supplier performance: 1) to select initially those suppliers to be used, and 2) to monitor and manage ongoing supplier performance levels.

Initial supplier screening and selection criteria can include:

- Supplier price, delivery, and quality performance reputation with their other customers. The names of noncompeting customers should be available as reference checks from potential suppliers about whom little is known.
- Whether or not the supplier is currently, or has plans to become, a Just-In-Time supplier. If so, what does being a Just-In-Time supplier mean to him/her, and is it a positive or negative experience?
- The financial stability of the supplier. 10K reports, financial statements, and recent issues of *Barons*, the *Wall Street Journal*, *Fortune*, *Forbes*, and *Business Week* are valuable resources for information on larger companies.
- Willingness to make long-term contractual commitments and to become involved in commodity-based agreements. (This is something that can be approached by pilot commodity and should always be addressed with much care.)
- Geographic location, including proximity and transportation availability.

Depending on how much these criteria narrow the field of supplier candidates, additional factors from the next listing may be added as well.

When the suppliers are in place, a number of aspects of their performance may be reviewed on an ongoing basis. Among these are:

- Quality performance over time–measured via incoming material inspection initially, then evolving into a combination of the suppliers' own records and service complaints as incoming inspection becomes superfluous.
- Delivery frequency and timing–its conduciveness to JIT operations, responsiveness to close-in changes, and delivery cost considerations; delivery reliability and willingness to do "point-of-use" delivery.
- Ongoing cost reduction activities–in the suppliers' operations, and with regard to long-term reductions in associated commodity prices to the manufacturer.
- Willingness of the suppliers to share their data with the manufacturer–and in a related vein, willingness to work on joint cost reduction/design improvement projects.
- Current ability–or planned development to provide capability in the areas of bar coding and electronic data interchange.
- Willingness to adopt special containerizing–to reduce cost and support JIT use of purchased materials by the manufacturer through reductions in material handling.

- Willingness of the supplier to maintain inventories at the manufacturer on a consignment basis–and to support consumption driven receipt/payment practices by the manufacturer.
- Levels of demonstrated cooperation–on warranty claims related to purchased parts and/or plan materials.
- Various aspects of the suppliers' operation, including:
  –Manufacturing lead time
  –Manufacturing capacity
  –Preventive maintenance activities
  –History and likelihood of organized labor disputes and interruptions in supply.

Supplier Certification

Closely related to sound supplier evaluation methods is the concept of supplier certification. When a supplier reaches "certified" status, it has been determined that it is no longer necessary to monitor the quality levels of incoming materials beyond an occasional random sample spot check. This reduces the cost of quality in the manufacturing organization incrementally as more and more suppliers are certified.

Buyer Performance Evaluation

The criteria for buyer performance evaluation frequently changes in a JIT environment. In traditional purchasing settings, buyer performance measurement has been predicated for a number of years on cost containment, with an occasional focus on delivery (availability) in the more progressive companies.

In Just-In-Time purchasing environments, the criteria are much broader in scope and more varied from company to company. The "big two" performance criteria (price and delivery) are still important in a JIT setting. In fact, delivery performance becomes increasingly important as buffers of work-in-process inventory and raw material diminish. Another performance measure as important as the "big two" is quality. The raw material received should consistently meet or exceed design specifications so that certification can occur, and incoming inspection is no longer required. By tying buyer performance measurement directly to the quality performance of his/her suppliers, you can promote responsibility and "buy-in" on the part of the buyer and constantly encourage him/her to strive for improvement and excellence in the supplier base.

Another important buyer performance measurement is the reduction of procurement lead times on his/her commodities. Responsiveness to customer needs and fluctuations in demand will ultimately be achieved only when the production cycle is minimized all the way upstream to the raw material in its original form. Adding procurement time reductions to buyer performance measures clearly states the importance of this principle and ensures its communication throughout the supplier community.

Supplier visits (both frequency and quality) are another important aspect of buyer performance. It is estimated that about 1% of buyer time is spent visiting and evaluating vendors (both potential and existing). Just-In-Time operations can most effectively be supported when buyers spend around 25% of their time face-to-face with suppliers and potential suppliers. Supplier evaluation should be done in a very structured, comprehensive fashion in these situations.

One of the most controversial buyer evaluation criteria that has been encountered is the improvement of supplier profitability, based on shared cost reduction. The concept involved here is that joint cost reduction efforts between manufacturer and supplier will generate cost reductions in purchased parts and raw materials. The cost reduction savings, then, should be split evenly between the supplier and manufacturer, with the profitability of both parties improving. However, this measurement is a real test of the attitude change in a manufacturing organization. An organization still clinging to the adversarial "we vs. they" vendor mind set will find the supplier productivity measurement very hard to swallow.

Other buyer performance measures that are used in different environments include continuing professional education efforts, early manufacturing involvement/value engineering activity, inventory level reduction, supplier base reduction, purchased part availability (service level), and variance from forecasted price levels.

Joint Cost Reduction

As described earlier, joint cost reduction efforts may be undertaken with the supplier in order to boost the profitability of both organizations. Typically, specific cost reduction targets are selected by a steering committee of both manufacturing and supplier management (often, this committee is comprised of materials managers and engineering managers). A team is selected of purchasing, quality, and engineering staff people. Training in problem solving, group dynamics, and cost/benefit analysis is provided, and the group is chartered with their objective and expected time frame for completion. They generally meet for about two hours per session, on a weekly basis. When cost savings are achieved and documented, they are reported to the steering committee. The reported savings are divided in half, and the purchase part price is reduced by this amount. In that way, savings are split between supplier and manufacturer. An added benefit to the supplier is that he/she is then able to transfer the new technology to other, similar product lines for other customers.

Supplier Training

It is vital that the supplier himself become a Just-In-Time manufacturer in order to realize the substantial cost and time reductions available in procurement. There are two ways to accomplish this: 1) use of the manufacturer's own people to train the supplier, and 2) use of outside consulting help.

Experience has shown that it is far easier to convince suppliers of the value of Just-In-Time operations once they have seen it work in their customers' factories. Simply asking suppliers to deliver what you want, when you want it, is asking for trouble. Suppliers must be shown the benefits of JIT operations, and trained in implementation techniques. This can be done with the manufacturer's personnel, provided that the manufacturer is sufficiently committed to the endeavor of providing training resources.

The supplier should be provided with training in each of the areas displayed on the "emerald" model, with examples developed from the supplier's own environment. Follow-up assistance should be provided as needed when questions and difficulties arise, and a "user-group" approach may be utilized to ensure continuous improvement.

Revisions to the Purchasing Organization

As the manufacturing environment evolves toward a more process-oriented flow, focus factories (or factors within a factory), as described earlier, will likely emerge. In this setting it may be advantageous to organize purchasing in such a manner that liaison materials people are assigned to each focus factory. While purchasing remains centralized, along with its contract negotiation and vendor-/market intelligence functions, the actual release of P.O. line items or kanbans and any required expediting/trouble-shooting occurs at the focus factory level via the liaison people.

In addition, depending on the level of involvement and political orientation of the rest of the organization, purchasing may assume a leadership role in multidisciplinary groups for purposes of vendor evaluation and cost reduction activities. Some organizations decide to incorporate engineering and quality control people into the purchasing department for these purposes.

IMPROVED DESIGN ENGINEERING METHODS

8

As mentioned earlier in this text, application of JIT techniques in the Design Engineering area falls into two categories: 1) Process and 2) Product.

Process-oriented improvements in Design Engineering include:

- Elimination of nonvalue-adding activities.
- Building manufacturability into the design.
- Building quality into the design.

Elimination of Nonvalue-Adding Activities

Over the years, it has become obvious that JIT techniques can be applied very successfully in office settings to reduce waste and improve productivity. Design engineering is certainly no exception. Beyond the obvious time-saving technological advances of Computer-Aided Design (CAD), Computer-Aided Engineering (CAE), and Computer-Aided Manufacturing (CAM), there is a host of applications of "natural" (as opposed to artificial) intelligence to streamline, simplify, and accelerate design and design change throughput. For example:

- *Reduce the number of Engineering Change Notices (ECNs) required*–This may mean adopting "quick change" forms for the simpler changes that do not need to follow the prescribed processing route. Another avenue to pursue is standardization of components, so that fewer changes are required. Finally, minor changes can be grouped, and/or put under "cover" change orders, with individual line item releases (similar to blanket purchase orders).
- *Improve the physical proximity of the processors of ECNs*–Putting together interdisciplinary "cells" of ECN processing people can provide important benefits in terms of communication and visibility. Like improved manufacturing flow, it also provides the ability to reduce defects in work-in-process ECNs by catching them immediately at subsequent processing operations.
- *Simplify and standardize ECNs*–so that they are "foolproof," and may be processed quickly without errors. Forms may be formatted and structured in such a manner that their completion and processing may be simplified and made more consistent.
- *Reduce the number of required approvals*–Combine approvals into a single, brief daily meeting, or simply designate any ECN involving less than a fixed-dollar amount to be exempt from the approval process.
- *Perform an evaluation of the process*–Identify bottlenecks, redundant activities, and unnecessary activities, and purge them. This is typically done with flow charts or data flow diagrams.

Building Manufacturability into the Design

The value of the design for manufacture (or design for assembly) concept is, like that of many other JIT principles, difficult to overstate. Because design engineering is such a complex area, it would not be reasonable to try to deal with it in depth here. However, there are some general guidelines for designing for manufacture (or assembly) that can be stated fairly succinctly, including:

- *Recognizing critical specifications and tolerances*–or, perhaps, recognizing which specifications and tolerances are critical and their degree of criticality. Many production lines have stopped and many parts have been reworked unnecessarily because specifications were unnecessarily tight. Realism is the key. In some instances manufacturing supervision "go to war" with design engineering departments over tolerances that, in the end, were almost meaningless, while elements that were critical to the customer were ignored by both parties. One such example (from the author's experience) involves construction equipment frame components that were specified to be true from a burn table operation within 1/16 of an inch over edge length of about 15 feet. The metal involved was 30mm thick and subject to significant movement from heat, shrinkage, etc. The spec would certainly not be a vital element in the end product, since additional operations (welding, shot-blast, etc.) would repeatedly alter the final result, and the customer had no such requirement in any event.
- *Recognizing existing process capabilities*–There are some very poignant examples of designs that were not able to be produced in situations where design engineers were physically distant from the actual manufacturing process. One organization constructed a beautiful new design engineering facility over four miles away from the manufacturing site, thus largely catering to a kind of "prima donna" whim and various political considerations. The resulting situation was predictably difficult with liaison engineering positions, hourly van service between the facilities, added design change time requirements, and seriously detrimental psychological "distancing" arising in the aftermath. Even in single-site scenarios, it is difficult to keep design engineers abreast of existing process capability. Machinery is replaced, modified, or removed with increasing frequency, and advertised equipment capabilities do not always match end results.
- *Recognizing similarity in parts and/or routings*–This is one area where recent advances in terms of computerized classification and coding (C and C), group technology, (GT), and computer-aided manufacturing (CAM) have helped tremendously. C and C will group parts according to characteristics (as described in an earlier section), making it easier to identify logical part families to be handled via group technology. Computer-aided manufacturing can allow the system to generate production routings, based on part characteristics and process capability data. While these systems are not perfect, engineering time can be substantially reduced through their utilization.
- *Recognizing existing "generic" routings*–A kind of "generic" routing is often devisable for part families or groups, based on the major processes and sequences of processes involved in the manufacture of the individual parts. Documented generic routings can ensure that critical sequences of production operations are maintained, while latitude is still available for unique part characteristics and their related production operations.

- *Designing to incorporate existing process capabilities and skill sets*–There is often a process termed "print review" in the adaptation cycle of new and revised parts. Typically, this involves a review of the part drawing, specifications, and characteristics by an interdisciplinary group. Members usually represent design engineering, purchasing, industrial and/or manufacturing engineering, purchasing, industrial and/or manufacturing engineering, and quality assurance. Manufacturing supervision representation is often useful in these processes, as well. It is important to keep these valuable sessions from becoming "rubber stamp" sessions that simply identify the parts as either purchased or manufactured. The parts should be considered in light of what major processes and process sequences are likely to be involved, and whether adequate process capability and skills exist to produce quality parts in an efficient manner. Machine capacity (current and projected) and product flow should also be considered, at least on a rough-cut basis.
- *Developing and utilizing preferred manufacturing practices*–The concept of preferred manufacturing practices centers around principles or "rules of thumb" that will facilitate production and may be incorporated into the design of the parts. Examples include minimizing the use of threaded fasteners, labels, soldering, and welding. Generally speaking, these principles reduce the amount of production/assembly time by replacing time- and/or labor-intensive practices with more efficient practices, tools, and devices. These principles should be jointly developed between manufacturing and design engineering and supported by manufacturing engineering, industrial engineering, purchasing, and quality control personnel as required. Beyond the obvious productivity advantages, there are tremendous psychological benefits from improved communications in this program.

Building Quality into the Design

As you may recall from the discussion about quality earlier, quality is best defined in terms of the needs (both real and perceived) of the customer. Customers can be internal (downstream machine operators, shipping people, etc.) or external (as in the case of the end consumer of the goods or services provided). Often, the identification of customer needs and dissemination of quality criteria are incorporated into the company's operation through a program known as Quality Function Deployment (QFD).

QFD has been defined in its engineering context as an approach in which the objectives are to determine target quality levels for design, to identify critical needs for the application of engineering resources, to identify potential design conflicts, to link internal control points to customer needs, to establish critical product and process control limits, and to develop operator procedures/training to achieve and maintain the designated specs. QFD is not unique or new in concept or in objectives. However, the advantage of QFD is that it provides structure, discipline, and continuity. It typically utilizes a series of charts (product planning matrices, or counterpart characteristics charts) to force the systematic assignment of specs and processes that will achieve each identified customer need. Each characteristic is ranked by importance, and customer perception of the organization's achievement of desired specifications is tracked over time. Like any other improvement program, QFD requires consistent application and discipline to succeed.

The other major aspect of building quality into product design is the fail-safe method. In the design process, a fail-safe method involves constructing the design in such a way that it cannot be produced or assembled in the wrong manner. Consider *Figure 8-1*, which depicts an assembly example:

This kind of engineering effort pays dividends in a number of areas. According to a recent article in the *Harvard Business Review*, Design for Assembly (DFA) programs have yielded these kinds of benefits in U.S. manufacturing organizations:

- Part number reductions of 15%.
- Subassembly reductions of 35%.
- Operation reductions of 30%.
- Manual assembly time reductions of 26%.

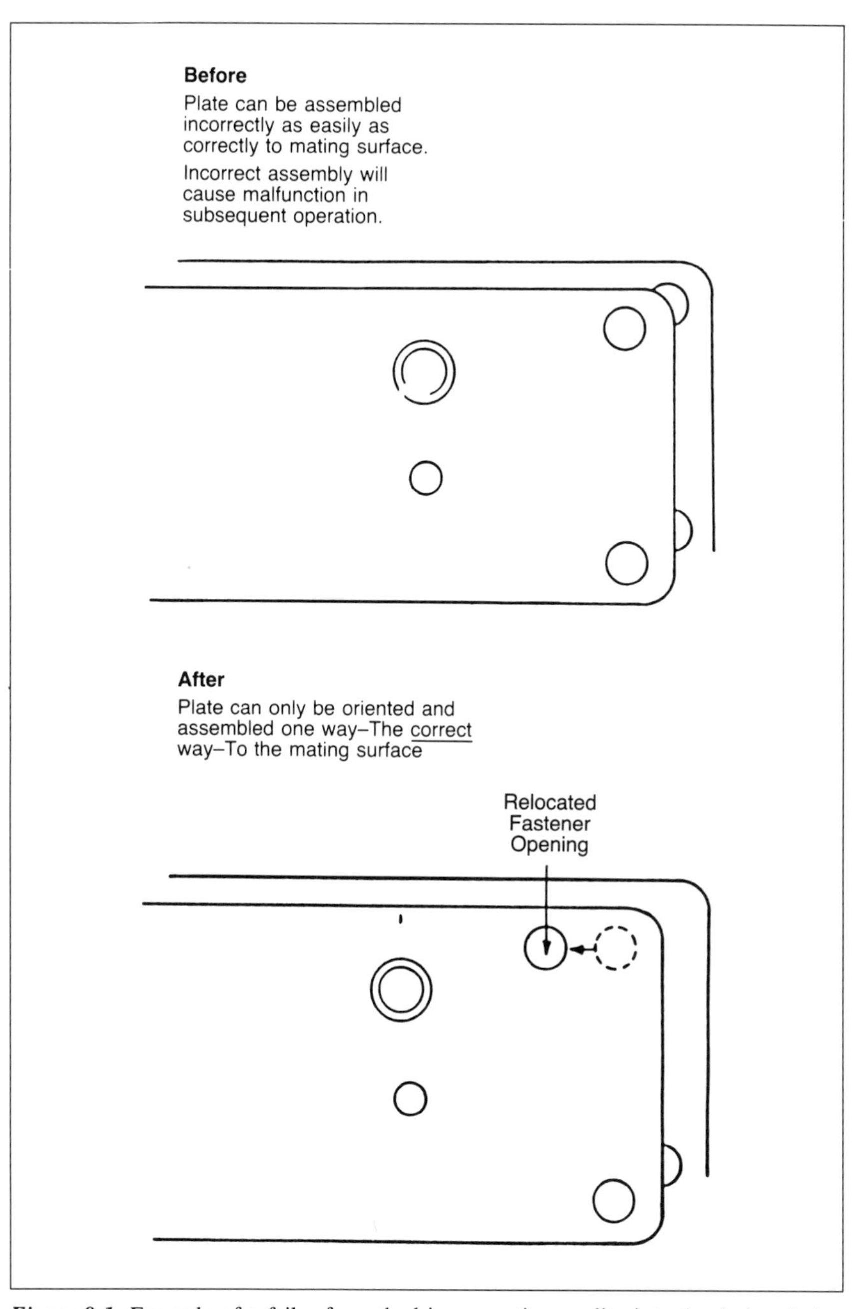

Figure 8-1. Example of a fail-safe method incorporating quality into the design during assembly.

ENHANCED SUPPORT FUNCTIONS 9

Applying the philosophy of waste elimination in support areas can be as challenging as any manufacturing setting. Every paperwork process, whether producing manufacturing routings or part costs, should be reviewed and streamlined. Many tools used to improve production operations can also prove valuable in the office.

First, the quality issue cannot be overlooked. Consider the ramifications of an incorrect effective date on an engineering change. Here are some potential impacts:

- Excessive obsolete inventory
- Warranty claims
- Incorrect routings
- Early or late raw material/purchased part availability
- Invalid customer promise dates
- Expediting/de-expediting of production orders (invalid due dates, start dates)
- Incorrect bills of material
- Inaccurate capacity planning
- Missed introduction dates on new products

Quality issues like these exist in every area and discipline. Poor quality is disruptive and extremely costly in every arena. Support areas are no exception. In fact, a particularly challenging area in terms of resolving quality problems is MIS. Imagine the consequences of software "bugs" in the various operating systems of a manufacturing or service organization. Not only are the programs themselves in need of problem identification and resolution, but output from the programs is suspect over an often indistinguishable period of time, affecting operating decisions, planning, and other program data.

In simplified synchronous production (SSP), elements of uniform work loading, housekeeping, and even setup reduction have application in the operations of support areas. A former colleague often walked through the office areas of client organizations, assessing their efficiency based on the "queues" of WIP inventory (paperwork) in incoming and outgoing mailboxes. Backed up purchase orders, quotes, routings to be processed, engineering change notices to be worked and memos to be read and/or filed can all be symptoms of an office area where work load is not evenly balanced between processors. If data entry work is stacked up because it takes so long to "log on" to the system that it is only done twice a day, there is a setup problem. When critical data is lost in a sea of paperwork, or in stacks of material on someone's desk, you have identified a housekeeping issue. Information and paperwork (whether it is bills of material, routings, accounting data, etc.) should flow as smoothly and efficiently as production material. The principles of SSP are an effective tool when religiously applied to accomplish this purpose.

We touched on the subject of applying process oriented flow in the office environment earlier in this text, using the example of constructing an interdisciplinary "cell" of engineering change notice processors positioned closely together to enhance communication and processing speed. Cellular manufacturing works well with paperwork processing just as it does with parts in the shop. In fact, when designing office floor layouts, moving away from the "functional area" layout with distinct, compartmentalized physical departments to a process orientation can be beneficial. Information types, like product lines in the manufacturing setting, must be identified and analyzed for volume and processing time at each step. Non-value-added activities are purged, and value-adding functions are aligned in a manner that will minimize redundancy, travel distances, and "WIP" inventories of data. Rather than moving the paperwork from department to department, people involved in processing purchase orders can be placed near each other.

Similarly, processing "cells" are created for engineering change notices, bills of material, customer orders, invoices, and/or any other data flow involving more than one department.

Advanced procurement technology is applied in the support areas in the following ways:

It is possible to regard the departments providing information as suppliers. The quality of their work should be monitored, as should delivery performance. When there is a problem in one or more of these areas, it should be quickly identified and corrected, rather than waiting for a downstream information processor to catch it, and incur significant levels of "rework."

In addition, it is desirable in many instances to bring "suppliers" of information into the internal problem solving activities. Many support department problems, for example, have been resolved by explaining them to representatives of MIS. Also, other departments may have experienced similar problems and developed a solution which can be transferred.

The most poignant example of improved design method applications in support areas is the design of forms. Simplification and ease of use should be the primary concern. Time and money is wasted by unnecessary documentation, compilation, and maintenance of data. Even when only "value added" data is utilized, the forms used are frequently so complex as to be confusing.

Beyond information processing, more fundamental and philosophical issues in many support areas may arise with the advent of Just-In-Time. An example follows.

In accounting, there are philosophical changes incurred in terms of cost accounting and accounts payable. In accounts payable, the potential exists for the elimination of most purchase orders, associated invoices and receiving transactions. Electronic data interchange can transform paperwork to electronic data. Beyond elimination of paper, the need for releases to P.O.s can be eliminated with the use of Kanbans (empty containers, empty trucks, etc.), and invoices can be eliminated through an agreement to pay based on production of finished goods. One finished product equals the receipt and use of 35 connectors, 18 fasteners, etc. Therefore, finished goods shipments can be multiplied by the bill of materials component quantities, and payment can be issued based on this calculation. This kind of system requires trust between manufacturers and suppliers, and internal departments at the manufacturing facility. It is often more difficult to overcome the mind set of the individuals involved than to work out the logistics of the system.

Cost accounting in JIT environments can also involve significant philosophical change. The costing process becomes more of a "process costing" or "activity-based" function than a discrete cost element roll-up. Work-in-process inventory virtually disappears, and so costing it becomes almost irrelevant. Stock issue transactions disappear, bill of material structures flatten out, and rework/scrap levels are vastly reduced.

In Production Planning, the philosophical changes involve master scheduling, production order generation, and allowances for scrap. Master production schedules (MPS) are usually still run. However, the schedule may be left at a relatively high level for purposes of capacity planning, and forecasted capacity requirements. Some JIT organizations never complete the MRP explosion at all from MPS. They merely provide suppliers with rough production requirements from MPS, recognizing they are buying supplier capacity, and final schedules will vary anyway as they get closer to actual production.

Production orders simply go away in some JIT environments, with final assembly schedules pulling Kanbans filtering back through entire production operations, and extending even to suppliers. Where production orders are still utilized, they are often relegated to prioritizing final assembly quantities and sequences. As long as pull systems are utilized, the first-in, first-out (FIFO) nature of operation sequence allows management to simply complete production orders and close them out in order as finished goods become available.

In conjunction with this, scrap allowances often go away entirely, and lot sizes get smaller, with an ideal lot size of 1. When scrap allowances disappear, it is critical that quality levels be high, and accountability for scrap be accurate and timely. Otherwise, the "pull" mechanisms employed will be desynchronized, and production will suffer.

Production control on the shop floor can practically be made obsolete, as final-assembly-driven pull systems eliminate the need for shop production schedules and expediting/de-expediting activities.

One area which becomes all the more vital in a JIT operation is Industrial Engineering. Balancing work loads from operation to operation to achieve uniform work loading, along with documenting and maintaining methods for production and/or assembly will be essential to smooth operations. In small job shops these activities might be performed by shop floor supervisors. However, in this setting the organization is heavily reliant on in-house expertise which tends to be unavailable at the most inconvenient times.

Another area ripe for waste elimination is distribution. Looking at each material handling and transportation activity to reduce non-value-adding occurrences can produce startling results. Beyond the elimination of non-value-adding activities, the areas of packaging, distribution channels, and markets may be addressed. Pareto analyses to study market penetration practicality and distribution channel and method effectiveness, often identify significant cost reduction opportunities. In addition, packaging can frequently be made more efficient via value analysis and value engineering techniques.

Another example of how JIT affects the accounting process was provided by authors William G. Holbrook and Robert Eiler. The two presented their ideas at the SME Synergy '86 Conference in a paper entitled *Accounting Changes Required for Just-In-Time Production*. This paper is located in Chapter 12.

EMPLOYEE INVOLVEMENT 10

A couple of years ago, I was involved in a Just-In-Time program with a manufacturer of houseware items. Looking at a specific product on the shop floor, I wondered how a particular process might be simplified. The machine operator, at whose work center I stood nearest, took interest in my activities. The operator shut off his machine, shoved his Chicago Cubs baseball hat back a bit, turned to me and said: "I know what you want."

"You do?"

He replied, "You want to do this all at one time? Come, I'll show you." I followed him to another machine.

To my amazement, he changed the machine around and made it run in reverse. Repositioning his safety glasses, he turned the machine on and completed the process in a single stroke. After examining the completed part, I said, "I don't understand. You just eliminated an entire production operation. Why haven't you told anyone about this before?"

His reply said more about the weakness of American management in four words than all the seminars I have ever attended.

He shrugged, smiled, and replied sheepishly, "They never ask me."

American management *must* learn to *ask* its employees for help. Over 70% of all valuable, implementable solutions come from the workers, both shop floor and office, not from management. Asking for help implies that management recognizes the value of other people's ideas. No one has all the answers. However, it is often difficult for management to accept that people lower on the organization chart than they are could have ideas they themselves had not considered. This is particularly evident in middle management ranks, where young, highly educated managers and supervisors oversee the activities of less educated but more experienced personnel.

The objective of an Employee Involvement (EI) program is to tap the resource of nonmanager's ideas, energies, creativeness, and innovation to solve operating problems and ensure continuous improvement. EI differs from Quality Circles and other similar programs in that EI groups are neither permanent, nor indigenous to a specific location. They may attack problems not directly related to quality, and report to an oversight or steering committee (*Figure 10-1*).

The EI organization is typically structured around the continuous operation of three to five task groups, whose activities are guided and monitored by an EI facilitator. The facilitator's responsibilities include EI group training, coordinating group requests for resources, assisting with group presentations to the Steering Committee, and assisting in regular meetings. The facilitator's role can be part-time or full-time, depending on the number of groups and levels of activity involved.

The sequence of activities involved in the operation of an EI program may be visualized as shown in *Figure 10-2*.

The process begins with employees submitting "suggestions" or "problem sheets," on which they describe the problems which give them the most

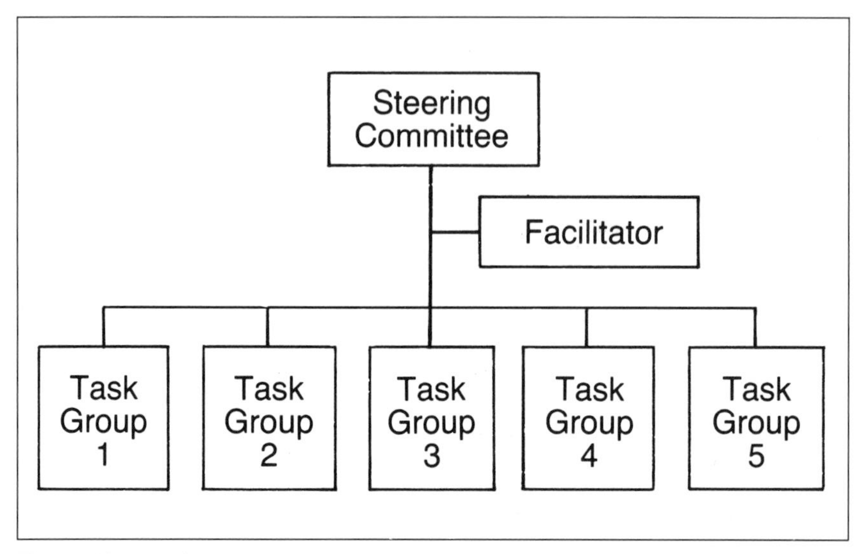

Figure 10-1. Typical Employee Involvement Organization.

difficulty in performing their day-to-day activities. Local supervision personnel are allowed to submit ideas as well. The sheets are accumulated and prioritized on a regular basis by the EI Steering Committee. The Committee selects the most critical problems and establishes a projected completion date for resolution of the problem. A problem solving task group of five to eight people is constructed, and the group is given its charter by the Committee.

The task group begins by organizing itself.

Regular meeting times are established, and a team leader is chosen with a recording secretary, who will take the minutes at all meetings. The problem is reviewed as it was described by the Committee, and any required added definition is obtained with the assistance of the EI facilitator.

When the problem is clearly understood by all task group members, cause-and-effect or problem solving activities are undertaken to identify potential underlying causes. There are a number of available approaches to problem solving in use in this area today. The fishbone cause-and-effect diagram, for example, helps identify and summarize a list of potential underlying problems, and prioritize them according to importance. Applying Pareto's Law, the list is narrowed to two or three potential causes, and data gathering is initiated to determine which really are the culprits. Depending on the nature of the potential cause, any of a number of data gathering tools may be employed, including control charts, interviews, sampling, designed experiments, etc.

When the underlying cause or causes have been identified, brainstorming is again undertaken to identify solutions. Solutions are prioritized, and evaluated for workability and implementation cost. A cost/benefit analysis is performed and the best solution is then selected.

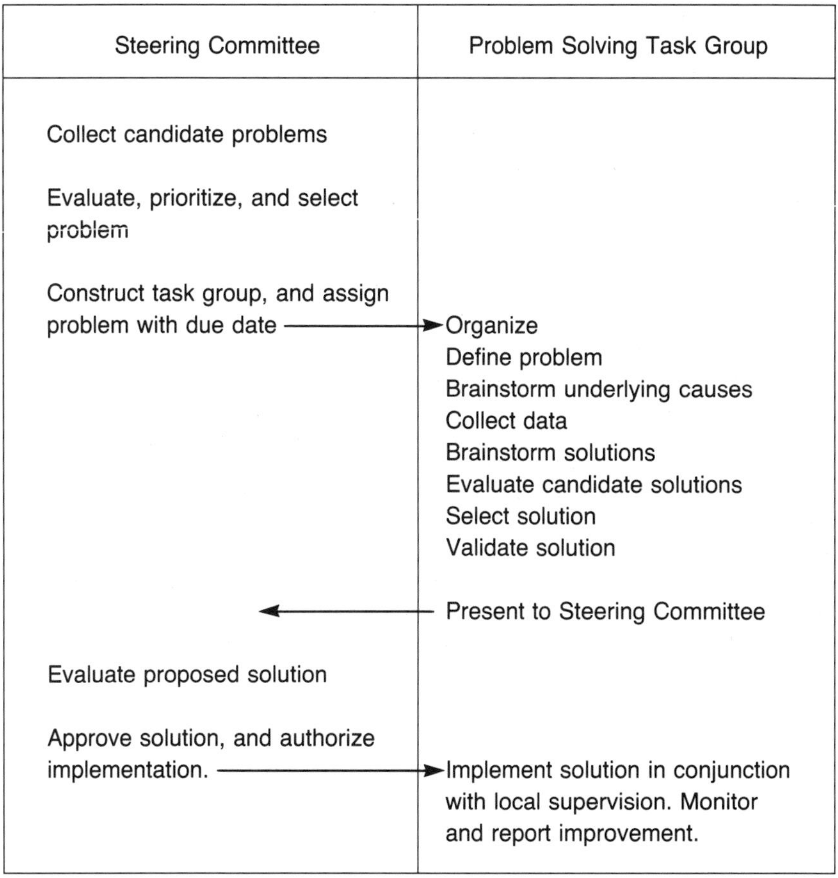

Steering Committee	Problem Solving Task Group
Collect candidate problems	
Evaluate, prioritize, and select problem	
Construct task group, and assign problem with due date →	Organize Define problem Brainstorm underlying causes Collect data Brainstorm solutions Evaluate candidate solutions Select solution Validate solution
←	Present to Steering Committee
Evaluate proposed solution	
Approve solution, and authorize implementation. →	Implement solution in conjunction with local supervision. Monitor and report improvement.

Figure 10-2. Typical Employee Involvement Operations.

After that, the facilitator assists the task group in preparing a presentation for the Committee. The team makes the presentation, answers any questions that the Committee may have, and generally receives approval at the conclusion of the meeting. The facilitator should be meeting with the Committee on a regular basis during these activities, keeping them abreast of the direction the task groups are taking. Accordingly, recommendations presented to the Committee should not be surprises, and approval can usually be handled promptly.

After the Committee has approved the recommendation, they assign an implementation team to report on the completion of the installation. Implementations should be conducted in cooperation with affected supervisors to enhance buy-in and reduce disruption.

When the installation is complete, the original task group (or a subset of that group) is assigned to monitor and report progress over a designated time frame (frequently three to six months). The Committee disbands the task group as implementation is completed.

Task groups initially will require training in the following areas:

- Group Dynamics–where the principles of working on problems as a team are taught. Team building, consensus attainment, group leadership, and other concepts are covered.
- Problem Solving–where cause-and-effect analysis is taught. Techniques include problem definition, brainstorming, fishbone charting, and Pareto analysis.
- Cost/Benefit–where the principles of cost and benefit assessment are covered. Concepts covered may include make/buy analysis, future value of money, calculation of carrying costs, overhead allocation, etc.

More technical training may be required occasionally in available tooling and production technology, process control charting, design of experiments, and a number of other areas. It is the facilitator's responsibility to identify the need for training as it arises in each group, and obtain it in a timely manner.

EI programs are typically self-funding after about one year. Savings can be enormous, yet unpredictable. Some savings, like improved safety, improved morale, and better communications are difficult to quantify. Others, such as reduced setup/changeover times, reduced defect levels, and increased resource utilization, are less ambiguous. Measurable or not, they are all important. Tracing their impact to the bottom line is one of the accounting professions's great future challenges.

Part II

Just-In-Time Implementation

Part II

Just-In-Time Implementation

Implementation programs differ widely from company to company. In smaller organizations (less than $100 million in annual revenues), it can take as little as six weeks to get through the planning stages and initiate implementation activities. In huge, multinational companies with complex manufacturing processes and deep BOM structures, the author has seen these same planning activities extend over nine months in a single division.

The activities themselves will vary as well. Some organizations that already have an accurate picture of current and projected operations will require less data gathering to identify product lines, volumes, etc. Also, process industries often find that greater opportunities exist in procurement and support areas, while discrete manufacturers frequently realize the greatest benefit in redesigning manufacturing layout and minimizing setup times.

We have established then, that implementation programs will vary substantially in time, activity-level content, and in opportunity areas/levels. As we discuss the details of JIT program operation, it will be important to keep this in mind. Not every activity will be required at the level of detail shown in every organization. Correctly assessing where the biggest opportunities lie in order to efficiently deploy available resources will be critical to program success, and will ultimately identify proper program activity contents and priorities.

While the activity-level content varies, the macro level process required is the same for all organizations. Broken into seven primary activities, the process can be visualized using the model on the following page.

Most of the balance of this text will be devoted to the components of this model, and their activity elements.

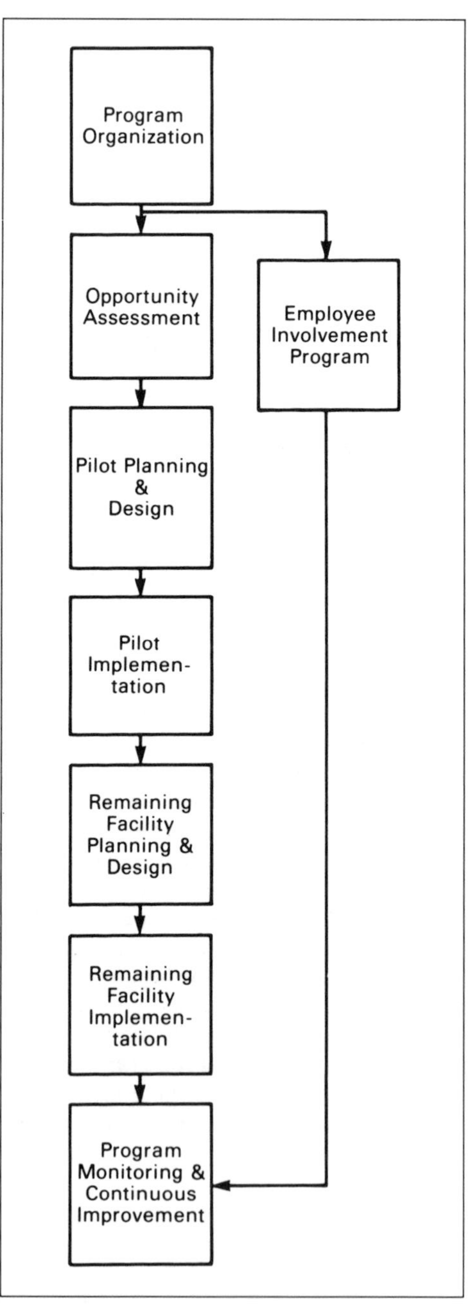

The organization of a JIT implementation program.

PROGRAM ORGANIZATION

11

As mentioned earlier, Program Organization is comprised of five primary components: 1) Awareness and Education, 2) Steering Committee Organization, 3) Opportunity Team Organization, 4) Employee Involvement (EI) Program Organization, and 5) Strategic Business Assessment.

Awareness and Education

Although often used interchangeably, awareness and education are significantly different. In an awareness session, the fundamental concepts of Just-In-Time philosphy and operations are covered. The presentation generally takes about four hours, and it is frequently supported by video tape, 35mm slides, and/or overhead transparencies. The objectives of the awareness session are to "sell" the concept and provide a common understanding of what JIT is. Laying this foundation with each and every employee in this fashion is a critical step in the overall success of the program. Achieving buy-in at the earliest possible moment provides enormous advantages, particularly when there is a labor union involved.

One of the best JIT implementations that comes to mind began with a two-day awareness and education session in the local VFW hall. All the employees (wage and salaried, including the janitors and the nursing staff) were in attendance, and the factory was closed for two days. The owner of the company began the meeting with a rather unusual statement. He stood, scanned the crowd with authoritative eyes until there was absolute silence, and said, "We're here today because I don't think you folks make enough money." He certainly had their attention. At this point, he went on to explain that they were all going to learn how to make more money for the company. As the company made more, they would make more. In addition, he explained that no one would lose his/her job as a result of the JIT program. Instead, as efficiencies were gained, additional work in the form of new product lines would be brought in. The content of his speech was perfect, and his delivery was masterful. But most importantly, he was as good as his word. Tremendous efficiencies were gained, including reductions in both throughput time and inventory of over 90%. New product lines were adopted within a year, and no one lost his/her job. The old incentive system was replaced with one based on worker flexiblility and quality, and employees did make more money.

With this introduction, employees paid close attention to the description of the elements of JIT and how each component could be applied in specific environments. The employees believed it when they heard that their ideas were needed and that management wanted them to work smarter rather than harder.

As we've already discussed, awareness sessions should include everyone. Often, the best ideas for improvement come from people not directly involved in the process being improved. These people can be referred to as "ringers." One such example might include one clerk, one data entry operator, or similar administratively oriented individual in a setup reduction team. These ''ringers'' perform an invaluable service by asking the questions others might consider too obvious or ''dumb.'' However, the only way that everyone can become a contributor to their work enviornment is for them to be involved in the awareness and education sessions held by the company.

The awareness sessions should include the objective of Just-In-Time philosophy, the elements of JIT programs (quality, SSP, etc.), the historical benefits of JIT operations, and the general objectives of this particular undertaking.

Education sessions are more in-depth studies of the various elements of Just-In-Time including:

1. *In the area of Quality*–Detailed discussions of QFD, SPC, Cost of Quality, Design of Experiments, and other TQC issues are held. A good deal of time is spent on questions and answers in this and all of the other education sessions. Sample exercises are often included so that instructors may be certain of their student's level of comprehension.
2. *In the area of Simplified Synchronous Production (SSP)*–Education includes setup reduction methods, uniform work loading calculation, and general work center housekeeping issues. Exercises in setup reduction can often be performed by utilizing fairly generic models. One model that has proven very successful, and that is original to this text, is the use of the ''Play-Dough Fun Factory'' to simulate setup reduction on an extrusion machine. (Quality issues, housekeeping issues, and a number of other related topics may be demonstrated with this device as well). The concepts of quick-change tooling, ''a place for everything and everything in its place,'' reliable equipment, fail-safe critical elements, and internal/external operations are demonstrable through a little imagination and very little investment. In addition, the extrusion device is one that everyone seems able to relate to. It can be set up in any conference room and transported easily.

   Uniform work-loading calculations can, of course, be performed anywhere. Exercises usually take on more meaning to participants when they can be constructed to include the trappings of the environments in question. For example, in an assembly environment the exercise should involve the balancing of work centers like ''fabrication, subassembly, and final assembly,'' while fabrication shop exercises might inlcude ''cut, machine, weld, and paint.''
3. *In the Process-Oriented Flow area*–it makes sense to include a brief exercise concerning the advantages of a pull system. A good example of this is a videotaped experiment developed and produced by Hewlett-Packard in their Greely, Colorado division. ''Before'' and ''after'' results are tallied as inventory, rework, and throughput times are reduced in front of the viewer's very eyes. The construction of similar exercises, utilizing management and direct-labor people from the organization involved, can be an extremely powerful tool for educating the work force and promoting ''buy-in.'' Similarly, cellular manufacturing layouts (U-shaped cells) can

be constructed with relative ease that demonstrate improved communications and manning variability.

4. *In the area of Advanced Procurement Technology*–Education should include some medium level of detail in the following areas:
   - Commodity-level purchasing management–including commodity contracting, commodity-level buyer performance measurement, and commodity-level vendor performance measurment.
   - Supplier relations–including how to treat suppliers as an extension of your own business, how to make JIT attractive to suppliers, the advantages of a narrow supplier base, and how to help suppliers become JIT operations themselves.
   - Data sharing–including what kinds of data should be shared, what means are available to share data more efficiently and reliably, and what advantages there are to be gained by data sharing.
   - Supplier certification and general evaluation–including all aspects of importance to the organization (quality, delivery, etc.).
   - Joint ongoing improvement efforts with suppliers–how they are structured, how they operate, and how resulting benefits can be shared to benefit both parties.
5. *In terms of Improved Design Methods*–Education should encompass the following areas:
   - Design for manufacturing and/or design for assembly–how it's done, why it's done, and by whom.
   - Early manufacturing involvement, (EMI), current product review, (CPR), and/or value analysis/value engineering program details–These sessions should cover the purposes of these programs, how and by whom the programs are run, and the expected benefits of the program.
   - Design for quality including failsafing techniques and preferred manufacturing practices. (Exercises may be included here as well).
6. *In the area of Enhanced Support Functions*–Education can vary widely. From organization to organization, support areas and levels of efficiency differ substantially. In addition, opportunity levels, and therefore implementation priorities will be different. Organization structures are often unique; this may change the makeup of the education program in terms of who gets which kinds of education.

   Nonetheless, enhanced support function education should generally include these kinds of elements:
   - Accounting in the Just-In-Time environment–This area will require detailed discussions by individuals who specialize in this field. Cost accounting, accounts payable, and even receivable invoicing activities may be fundamentally changed and thus require thorough examination and understanding.
   - Production planning and inventory control–including discussions of how existing lot sizing, manufacturing order generation, and production

scheduling activities are likely to change.
- Distribution–including how transportation efficiencies may be realized and how waste can be drained from the distribution channels.
- Changes that may be required in MIS to support Just-In-Time operations–MIS in JIT environments will usually find that different kinds of data are required, that the data is aggregated and analyzed in different ways, and that it is needed in a much more timely manner.
- General paperwork processing, especially involving waste elimination activites in areas like customer order entry, purchase order processing, and design change notice processing–The primary benefit of education in this area is to alert administrative personnel of the opportunities in their areas.

Steering Committee Organization

The steering committee is organized for the purpose of overseeing planning and implementation activities. The primary functions of the committee are the review of ongoing planning activities on at least a semimonthly basis, approval of implementation plans prior to their execution, overseeing the actual implementation activities (again on at least a semimonthly basis), and approval (authorization) of all significant program-related expenditures.

The steering committee is typically comprised of senior facility-level management (plant manager, works manager, controller, and direct reports) and often includes representatives from any labor union(s) involved. (Outside consultants are often brought in to participate in steering committee meetings as well).

As the program evolves out of its focus on implementation and into the monitoring/continuous improvement phase, some subset of the steering committee becomes permanent for the purpose of directing employee involvement program activities.

Opportunity Assessment Team Organization

The assessment of opportunities for improvement is very critical to overall program success because it lays the foundation for all future program activities. Based on this assessment, priorities are determined by implementation, and limited resources are deployed in such a way as to maximize their use. It is essential that the assessment team understand the principles of JIT, the workings of the organization involved, and what reduction opportunities look like. For this reason, it is generally advised that the opportunity assessment team be comprised of both existing organization management personnel and experienced outside consultants. Often, especially in smaller organizations (less than $100 million in sales), the team is made up of two full-time individuals (a consultant and a mid-level manager), with technical resource assistance provided on a part-time basis as needed from other consulting and managerial staff.

The objectives of the assessment team are to identify existing levels of opportunity for waste elimination within the organization and to develop priorities for implementation that will both maximize return and establish a sound framework for future operations.

Employee Involvement (EI) Program Organization

The organization of an EI program, and program launch, can successfully be undertaken while other JIT program activities are underway. It will require that operating guidelines are constructed, that the steering committee is put into place, and that a facilitator is trained and assigned. Thereafter, employees will need to be made aware of the program and solicited for participation. (Experience indicates that voluntary participation is far more effective than mandatory employee involvement. Following the program's initial success, peer pressure and a feeling of "missing out on a good thing" will usually suffice to win over most holdouts and skeptics). Initial project topic suggestions are accepted for evaluation by the steering committee, and the program is underway.

The balance of EI program activities are described in the section entitled "Program Monitoring and Continuous Improvement."

Strategic Businesss Assessment

As part of a company's opportunity assessment, a strategy must be developed to effectively utilize capacity, equipment, and people that efficiency improvements will leave unused. This is not easily done in industries where the market is mature, or declining. It is even more difficult when the company involved is the only major player in the market. Forcing management to come to grips with these issues early on can minimize the pain of dealing with these circumstances after they occur. Analyzing potential new markets and product lines can stimulate a great deal of improvement in these and other areas as well, including revitalizing sometimes stagnant management teams. Often a company becomes dangerously reliant on a single industry, when there is sizable growth potential in other areas.

The point, then, of strategic business assessment in the early stages of a JIT program is to ensure that resulting efficiency and productivity improvements contribute to the growth—not the downsizing—of the company. Of course, this is not always possible but it usually is. The strategic assessment and its strategic plan often indicate whether company gains will be utilized to contribute to the growth of American manufacturing, or to its demise.

The strategic business assessment is comprised of an analysis of existing product lines, markets, technology and capability, customer base, and organization. It evaluates a number of factors in each category. For example:

- *Product lines*–including sales dollars, production volumes, and projected growth/decline levels by product line.
- *Markets*–including what industries/markets are currently served, what other industries/markets exist for current product lines and/or manufacturing capabilities, and the projected growth/decline of each.
- *Technology and capability*–including existing equipment capability and how it performs relative to design specifications, and to competitor

capability; technological position, relative to the "state of the art", and to competitors; and projected improvements in both internal and industry capability.

- *Customer base*–including an analysis of sales volume, profit margin, and shipment volume by customer, and of expected growth.
- *Organization*–including an analysis of the ability of existing work force to meet projected demand levels, of existing and planned levels of cross-training, of existing skill levels, and of cultural aspects of the organization that may inhibit or facilitate improvement.

The culmination, then, of the program organization phase of JIT implementation, is an environment that contains an educated management and labor work force, a functioning program steering committee, a functional opportunity assessment team, a developing EI program, and a completed or nearly completed strategic business assessment.

OPPORTUNITY ASSESSMENT 12

Opportunity assessment is designed to identify opportunity levels by area and product line, so improvement activities may be prioritized, and accurate benchmarks are established to measure future improvements. At this stage, employee involvement program projects are also begun.

The categories of opportunity assessment activities include: General Management, Quality, Simplified Synchronous Production, Process Oriented Flow, Advanced Procurement Technology, Improved Design Methods, Enhanced Support Functions, and Employee Involvement.

General Management

Opportunity assessment will require data gathering in the following areas, as it pertains to general management:

1. Identification of product lines defining product lines as groups of products or components which require the same manufacturing resources and/or raw materials for their production.
2. Identification of costs and profit margins associated with the product lines identified previously.
3. Identification of current and projected growth levels associated with the product lines.
4. Planned or forecasted changes in product mix and/or markets served which may affect the capacity and/or capabilities of the manufacturing organization.

(Much of this information should be available from the strategic business assessment).

Based on the data gathered, analyze product lines to determine a general course of action for each line. The analysis may take the form of the model (shown in *Figure 12-1*).

The model is divided into four quadrants, represented by the numbers 1-4, and distinguished by levels of production volume and profitability. Generally speaking, it seems wise to attempt to reduce cost on low profit margin lines, reduce throughput times on high production volume lines, improve quality levels on high profit margin lines, and increase sales on low production volume lines.

Placing each product line in a quadrant, based on the data gathered, it is possible to develop an overall strategy by product line. For example:

- Quadrant Number 1 product lines–those which are high volume but low margin, are usually excellent candidates for cost reduction programs such as Value Analysis, Design for Manufacture, and/or Material Cost Reduction.
- Quadrant Number 2 product lines–those with high volumes and high margins, are generally candidates for repetitive manufacturing techniques. These include process oriented flow, focus factory applications, and dedicated production equipment.

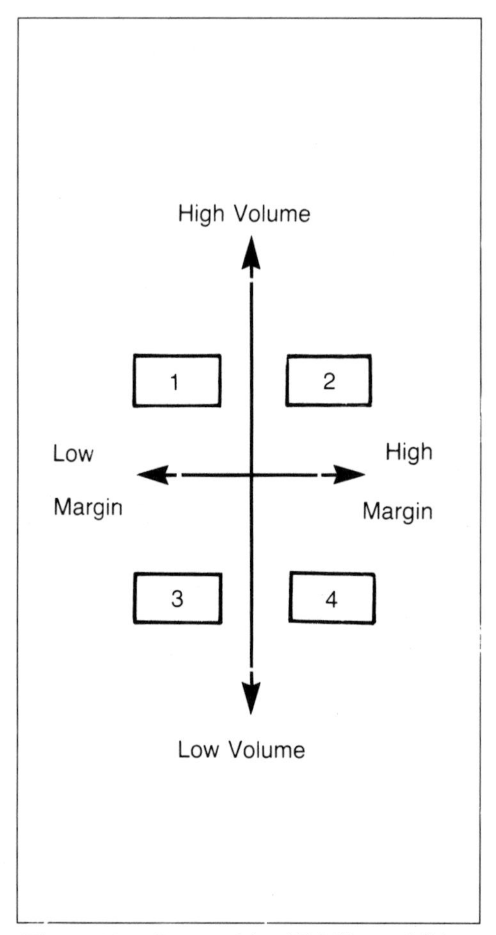

Figure 12-1. Product Line Viability and Disposition Model

- Quadrant Number 3 product lines–those with low volumes and low margins, are often considered for elimination. The questions which must be answered in such cases are "What are expected growth levels?", "What are expected cost reductions?", and "What would the impact be on customer?"
- Quadrant Number 4 product lines–those with high margins and low volume, are candidates for job shop or "process cell" applications. Flexible equipment and heavy operator cross-training are strongly recommended.

It will be important to retain the documented strategies for each product line, along with management's strategy for any new product lines and/or product mix changes to include in the overall improvement priority assessment.

Quality

The first item to be documented is the cost of quality (COQ). Identifying, documenting, and monitoring these costs is important, because:

- It allows further analysis to be performed.
- It focuses management and staff attention on COQ issues.
- It provides a benchmark against which to track improvement.

When COQ has been measured, the elements should be subjected to Pareto Analysis, and evaluated as to opportunity for resolution, and related anticipated savings.

Next, customers should be polled to identify and quantify their perception of critical quality characteristics. External customer needs should be reviewed first, then those of internal customers. Document all customer needs, and rank them in order of importance. Determine whether these characteristics are reflected in product specifications, and note any which are not. Working with manufacturing and engineering departments, determine which processes produce the "customer-critical" quality characteristics. Identify whether the process involved are capable of consistantly producing to the stated specs, and note any which are not. Track those characteristics defined as "customer-critical" over a significant period of production, to identify "defect" rates. Work with the Quality Control and Marketing deparmtents to assess or estimate existing and/or potential loss of sales from failure to meet "customer-critical" specifications.

In addition, review all characteristics deemed "critical" in current design specifications. Investigate whether process capability exists to produce them, and how consistently they are being produced within spec in the current environment. Working with Quality Control and Manufacturing, assess the opportunity for improvement, based on the application of SPC, Design of Experiments, and/or failsafing techniques. Where possible, quantify (dollarize) the potential savings resulting from these improvements.

Utilizing a combined listing of "customer-critical" and "design-critical" processes and characteristics, work with Quality Control and Manufacturing to determine most significant requirements for specialized operator knowledge. Review existing operator capability levels to determine critical training needs, and document them.

Finally, working with manufacturing and maintenance, identify the percentage of downtime associated with existing equipment over the previous 12 months. Analyze the percentage of downtime which may have been avoidable through preventive maintenance, and what costs may be attributed to the loss of manufacturing capacity and/or capability over those periods.

Simplified Synchronous Production

The data which needs to be gathered in the area of SSP falls into these categories:

- Current and projected demand levels–by product line and by machine or cell.
- "Standard Hours" listing–by work center and part number.

- Existing throughput times.
- Individual machine/work center/cell cycle times.
- Existing equipment breakdown and maintenance history.
- Existing equipment capability levels.
- Existing setup/changeover times–for all production equipment.
- Levels of existing manpower interchangeability.
- Existing data on housekeeping and safety.

Based on this data, identify existing and potential bottleneck machines/work centers/cells, by applying projected demand to current capacity. This should be done in several ways. First, by using existing measures of traditional "standard hours." This measurement will assume existing lot sizing practices and operating efficiency/utilization levels. Second, by using a lot size of one. Third, by assuming existing lot sizes but setup times reduced by 75%. Fourth, assuming a lot size of one and setups reduced by 75%. Using these findings, assess improvement opportunities available through setup reduction and reduced lot sizes. Remember, one-piece lot sizes are difficult to achieve, and you are unlikely to reduce all setup/changeover times by 75% in the near term. ("near term" generally equates to six to nine months.) Fifty percent average reductions are reasonable in the near term, and 75% is an obtainable target over longer periods. Accordingly, each assessment team must reach an estimate based on their own knowledge of the operation, and of results in similar cases.

Work with manufacturing, manufacturing engineering and/or industrial engineering to estimate potential productivity improvement levels available through cross-training and physical proximity changes, to enhance operator interchangeability.

Next, work with manufacturing and related engineering functions to analyze existing and forthcoming equipment, and develop an overall equipment strategy. A model used successfully for this purpose is shown in *Figure 12-2*.

Similar to the product line assessment model, this model is divided into four quadrants based upon the characteristics of capability and utilization. Capability and utilization may be defined differently for purposes of analysis, depending on the circumstances. It is useful to define capability as "ability to consistently produce parts to design specifications". In utilization, while the scale is different in virtually every organization, it is generally advisable to begin by assuming that "low" is zero to 49% utilized, and "high" is 50 to 100%.

Each piece of equipment is rated jointly by manufacturing and the appropriate engineering departments, and placed in its designated quadrant in the model. Various strategies can then be developed for the equipment, based on the balance of capability against current and future need.

- Quadrant Number 1–is characterized by low capability and high utilization. Typical strategies for Quadrant number 1 equipment include loosening of design specs to meet equipment capability, off-routing work to other work centers and eliminating the equipment, or replacing the equipment with more capable equipment.
- Quadrant Number 2–is characterized by high utilization and high levels of capability. Typical strategies applied to equipment in Quadrant number 2

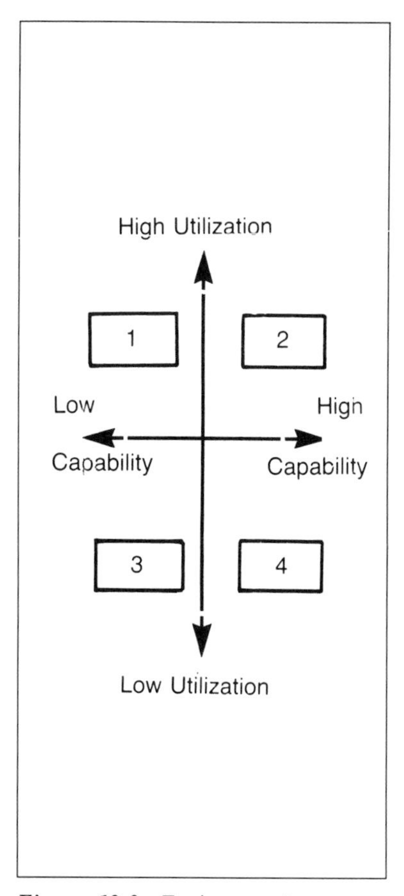

Figure 12-2. Equipment Assessment Model.

include duplication of the existing equipment as needed to meet demand, setup and changeover time reduction to increase availability of productive time, and sound preventive maintenance activities to minimize downtime.

- Quadrant Number 3–is characterized by low levels of both capability and utilization. Typical strategies utilized with Quadrant number 3 equipment include loosening design specs to meet machine capability, routing more work over the equipment, refurbishing equipment to improve capability, and off-routing existing work to eliminate (sell) equipment.
- Quadrant Number 4–is characterized by high levels of capability and low utilization. Typical strategies applied to equipment in Quadrant number include routing additional work over the equipment to improve utilization, bringing in currently farmed-out or new work to run on the equipment, and sale of the equipment.

Finally, identify and document existing levels of housekeeping performance. Frequently, no such data exists, and must be generated by audits of the manufacturing facility. The data should be as objective as possible, and specifically measurable. Measurements may include number of machines with visible oil leaks divided by total number of machines, number of machines with trash or offal on floor divided by total number of machines, and zone-by-zone ratings of number of occurrences of materials in designated aisles. Work with manufacturing to determine time allocated and actually used for housekeeping activities, and the number of reported safety incidents related to housekeeping issues. Based on this, estimate potential savings opportunities in the area of housekeeping to be achieved by not allowing trash, offal, etc. to hit the floor in the first place, and by preventing housekeeping-related safety incidents.

Process Oriented Flow

The most important data to be gathered in Process Oriented Flow involve the documentation of current product flow, material handling points, travel distances, and a queue analysis.

The documentation of product flow, including travel distance, inventory level, and value-added levels, can be easily accomplished through use of the value-added analysis. It involves physically tracking the flow of all major product lines on the shop floor, carefully noting the materials' path and each point of material handling on a scale drawing of the outline of the plant.

This should make it possible to analyze product flow and material handling efficiencies in the following ways:

- Divide the number of value-adding operations by the number of total operations to obtain the percentage of material handling occurrences which add value.
- Add individual travel distances to identify total travel distance by product line.
- Identify bottleneck operations by the consistency with which inventory stacks up ahead of individual operations. As important as how much inventory exists and where, is the question of why. All WIP inventory documented during the value-added analysis should be identified on a separate sheet, converted to dollars, converted to days of production, and assigned a reason code. It should then be possible to identify those reasons most heavily contributing to Work-In-Process.

An additional aspect to be evaluated is existing throughput times. There are a number of ways of doing it. The most straightforward is simple observation, but it requires weeks or months to perform an accurate analysis of all major product lines this way. Another is the "cycle time analysis," which relies heavily on production reporting data. Usually, a simple record of start dates and completion dates as reported to the Shop Floor Control (SFC) system will do. The throughput times are depicted with alpha characters, each different character representing a different operation, with three characters representing three units of time.

This kind of analysis, especially applied over all major product lines, will allow the assessment team to home in on the relatively small number of operations which represent the majority of throughput time, and assess the opportunities for improvement in those cases.

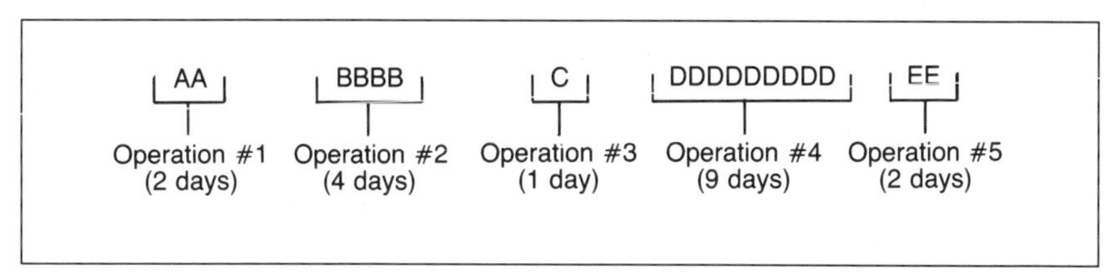

Figure 12-3. Cycle Time Analysis.

Advanced Procurement Technology

Data gathering and analysis in the category of Advanced Procurement Technology (APT) falls under these headings: Commodity Analysis, Supplier Base Analysis, Ordering Method Review, Joint Improvement Efforts Assessment, and a Buyer Time Analysis.

In commodity analysis, the assessment team usually works with the buyers to identify what commodities are currently purchased. (Non-JIT purchasing departments do not typically manage by nor in some cases, even identify commodities). As mentioned previously, a commodity is a group of purchased items which require the same raw materials and/or equipment for their production. In some cases, the definition may even be broadened to include more than one commodity which may be purchased from the same source, such as electrical connectors and wiring harnesses. A number of data elements should be evaluated with regard to the commodities currently purchased, including: number of part numbers per commodity, annual purchase volume in dollars and pieces, inventory turns, number of major (over 10% of the part volume) and minor (10% or less of the part volume) suppliers, and average lead times.

This information should make it possible to assess reducing the supplier base to one or two per commodity. This not only enhances the manufacturer's negotiating position in terms of cost, quality, and delivery requirements, it provides significant improvements in schedule change flexibility as well.

Supplier base analysis reviews existing supplier base to identify opportunities for reduction and improvement. Data gathered by supplier should include commodities provided by the supplier, the annual purchase volume (in dollars and pieces) attributable to the supplier, the percentage of the supplier's capacity represented by your annual purchases, the geographic distance of the supplier from your own facility, and the average lead time attributable to procurement of parts from the supplier. In addition, it will be important to identify the supplier's historic performance in the areas of price, quality, and delivery performance, and consider any contraints imposed by suppliers such as minimum order quantities and penalties for less-than-lead-time orders. Finally, you may wish to investigate the supplier's involvement with JIT techniques, and their willingness to become involved in your own JIT endeavors.

With this data, it should be possible to determine opportunities for reducing the supplier base, and achieving gains in quality/cost/delivery/lead time. A list of candidates may be developed for joint cost reduction efforts as well. Ordering methods reviews involve assessing volume, value, frequency, and methods of ordering purchased parts and materials. Order volumes can be decreased and flexibility can be gained using commodity based contracting, blanket orders, and intermittent releases against these long-term orders via kanban. By identifying existing order frequency and activities associated with the ordering process, it is possible to estimate the savings potential of JIT contracting principles. It is a matter of costing the order-related activities, and projecting the reduction in orders.

Since it is not possible to accurately estimate potential savings from joint cost reduction efforts with suppliers, evaluating these efforts is nearly binary. There is such a program or there is not. It is worth noting that savings generated from these programs, like other value engineering types of undertakings, can reduce costs of materials and/or parts by more than 50%. However, success varies depending on the personnel and resources committed.

Finally, it is usually a valuable exercise to perform a buyer time analysis. This can be done by using a time log broken into 15-minute increments, with buyers completing the log by recording an alpha code associated with an activity type (A for vendor visits, B for telephone calls, C for purchase order generation, etc.). The results of aggregating the percentages of buyer time activity and pie charting them, is shown in *Figure 12-4*.

The kind of picture we would like to see with the application of JIT principles is more like *Figure 12-5*.

The purpose of this exercise is to arrive at an estimate of improved buyer efficiency, with a substantial shift from ordering and paperwork related activities toward vendor visits and cost/lead time reduction pursuits.

Improved Design Methods

To assess opportunity levels associated with JIT in the Design area, the categories of analysis work include Engineering Change Notice (ECN) Review, Design Quality/Manufacturability Review, and a review of any existing value analysis/value engineering, CPR, and/or EMI programs.

The engineering change notice review involves identification of current and projected ECN volumes, thoughput times, and part number changes. The assessment team should also note whether blanket ECNs are utilized, and the degree of standardization in use. Once this is known, and after a brief analysis of queues, duplication and redundancy, and other non-value adding activity in the ECN process, it should be possible to estimate potential reductions in ECN throughput time, processing manpower requirements, and cost.

The review of current designs for manufacturability and quality will need to be done jointly between the Q.C. Department, Manufacturing Supervision, and the various engineering disciplines in conjunction with the assessment team.

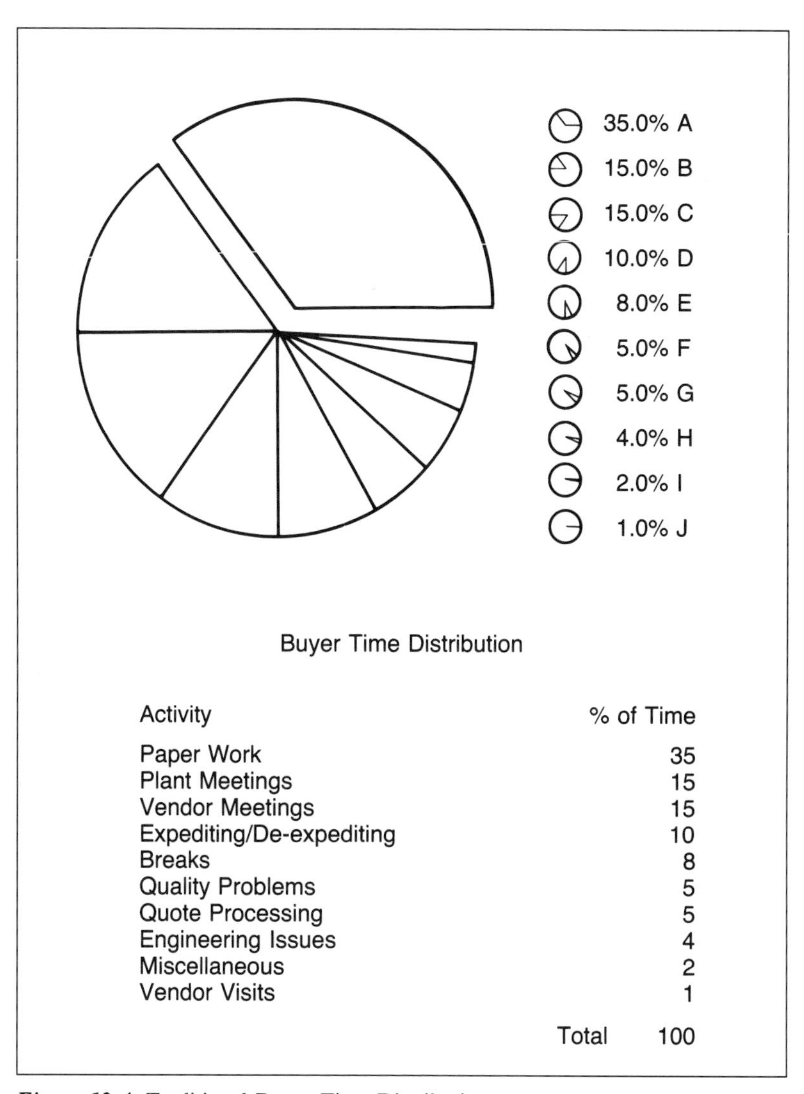

Activity	% of Time
Paper Work	35
Plant Meetings	15
Vendor Meetings	15
Expediting/De-expediting	10
Breaks	8
Quality Problems	5
Quote Processing	5
Engineering Issues	4
Miscellaneous	2
Vendor Visits	1
Total	100

Figure 12-4. Traditional Buyer Time Distribution.

In manufacturability, the primary concern is process complexity. Gaging current process complexity is both a subjective and objective process. Objectively, it is a mattter of quantifying the number of tools and operations involved in the production of the design, and identifying how close the design tolerances are to process capability limits. Subjectively, it becomes a "reasonableness" test: What is reasonable to expect from an operator and a piece of equipment under these circumstances, and in this environment? A work sheet may be

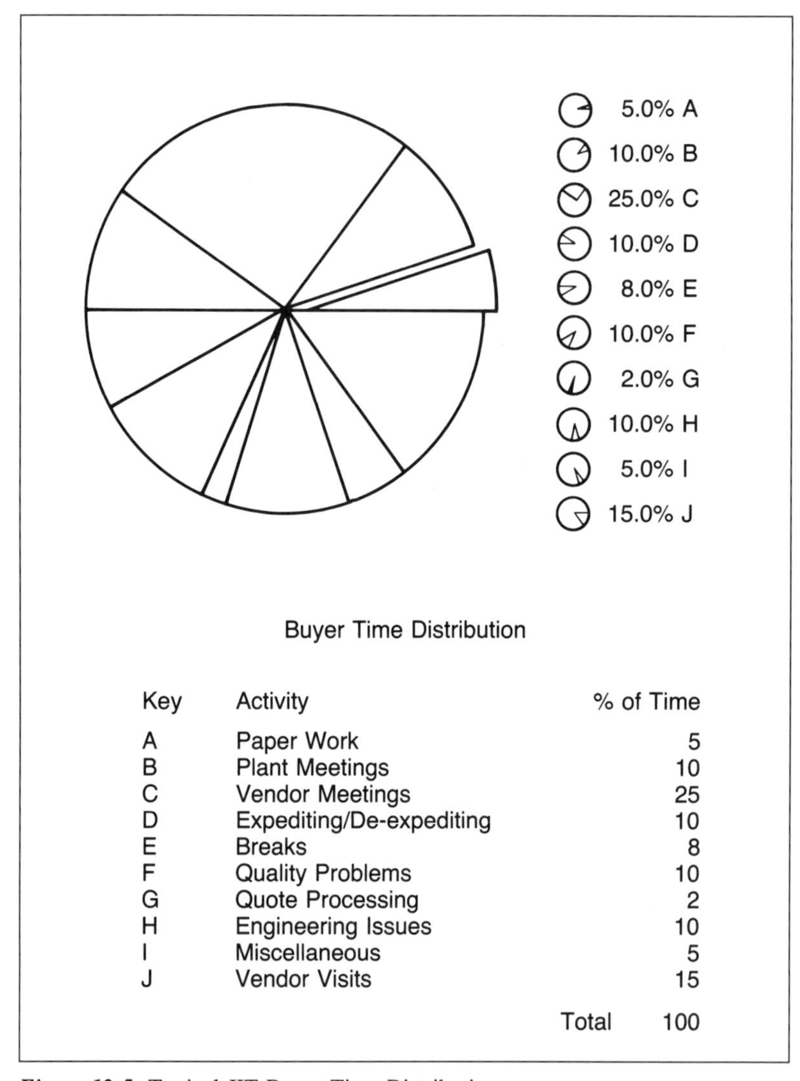

Key	Activity	% of Time
A	Paper Work	5
B	Plant Meetings	10
C	Vendor Meetings	25
D	Expediting/De-expediting	10
E	Breaks	8
F	Quality Problems	10
G	Quote Processing	2
H	Engineering Issues	10
I	Miscellaneous	5
J	Vendor Visits	15
	Total	100

Figure 12-5. Typical JIT Buyer Time Distribution.

developed from critical specifications lists and production routings to assess the opportunities in these areas.

The primary concerns with quality are:

1. Consistently meeting those specifications and characteristics deemed critical to product safety and reliability.
2. Consistently meeting the specifications identified as critical to customer-defined quality characteristics.

Generating designs to meet these criteria involves making the design "failsafe," so opportunities to mis-produce or mis-assemble are minimized.

Identifying opportunities in these areas involves documenting the current volume of defects (out-of-tolerance parts) which are attributable to:

- Unreasonable specifications
- Unnecessary specifications
- Unavailable process capability.

It is also necessary to document the percentage of throughput time and/or rework attributable to design complexity.

Finally, it is important to assess any ongoing programs in Value Analysis/ Value Engineering, Current Product Review, and/or Early Manufacturing Involvement. The number of designs reviewed and their complexity and value should be compared to the total design population. Future savings opportunity may be estimated based on historical savings achieved.

Enhanced Support Functions

The following departments will be discussed under Enhanced Support Functions:

Production Planning
Production and Inventory Control
Accounting
Distribution

In Production Planning, the data needed includes production order volumes, and the number of manhours attributed to generating, maintaining, and closing them. The assessment team can then apply the concept of generating a final assembly schedule from MPS, and driving assembly, subassembly, and fabrication operations via pull system. Recognizing this will be a phased-in approach applied one product line at a time, the assessment team can estimate the manhours saved by not generating, maintaining, and/or closing production orders in increasing volumes over time.

Production and Inventory Control opportunities may be found by identifying the manhours currently spent prioritizing, reprioritizing, expediting, and de-expediting on the shop floor and in "hot list" meetings. Applying pull system concepts, the need for these activities is eliminated.

Accounting opportunities may be assessed in the following categories: accounts payable, accounts receivable, and cost accounting. In each accounting area, the assessment team will need to document the number of manhours attributable to each daily activity. Then, in the payables area, opportunity may be estimated by considering what percentage of these activities may be eliminated through the application of:

1. Backflushing bills of materials–and feeding the payables system with the data (eliminating the need to base payables on invoicing).
2. Utilizing electronic funds transfer eliminating the need to perform the current check writing activities.

In the receivables area, opportunity levels can be discovered by estimating what percentages of manhours may be eliminated via electronic invoicing and electronic funds transfer. In terms of cost accounting, the assessment team will need to identify what percentage of manhours may be eliminated through the lack or need to value WIP inventory at various stages of completion, and a general reduction in all forms of inventory. Simplified routings and designs will contribute to more streamlined cost accounting, as will generic routings, and process cells.

Finally, in all areas of accounting it is the author's experience that there are significant opportunities to be found in flow charting the paper work flow, looking for:

- Transaction timing problems
- Redundant processing steps
- Unnecessary processing steps (approvals, etc.)
- Missed steps, or "holes" in the process.

When the paperwork processes have been flowed and opportunities have been identified, they must be documented and valued.

Another example of how JIT affects the accounting process was provided by authors William G. Holbrook and Robert Eiler. The two presented their ideas at the SME Synergy '86 Conference in a paper titles *Accounting Changes Required for Just-In-Time Production.*

(The italicized material in parentheses in this paper has been added by the author of this book. This material should offer additional information from this book's authors point of view).

(*As we have seen in Chapters 5 and 6*), Just-In-Time production requires flexibility on the shop floor; likewise, the accounting practices used in a Just-In-Time production environment must be flexible. A fixed set of costing routines will not always be applicable in this environment of constant change. Traditional reporting and cost methods require severe scrutiny, and current tools may become inadequate. For example, the emphasis traditionally placed on work-in-process inventory valuation will be altered because Just-In-Time concentrates on reducing inventory levels.

Some articles on Just-In-Time equate inventory level with the water level in a pond; problems that might occur on the shop floor are compared to rocks on the bottom of the pond. A high inventory level only hides the real problems (e.g., inefficient production methods) that often will surface when the company is least able to resolve them (e.g., in an economic downturn). The Just-In-Time philosophy forces the inventory level down, exposing the problems so that they can be solved.

The information provided by the accountant should help management identify the problems, determine their significance, and aid in their ultimate solution. The changing role of the accountant necessitates changes in traditional accounting functions. This article discusses some of the internal changes that are required as companies move toward Just-In-Time production. In addition, the external accountant will need to develop new approaches and techniques to be consistent with this new manufacturing environment.

The Financial Reporting System

(*As with the other disciplines depicted in Chapter 1*), financial reporting systems typically evolve and mature over several decades. In some companies, these systems have become so complex that the reports they generate often are virtually useless to manufacturing management. The Just-In-Time environment, in which problems must be allowed to surface, uses a simplified reporting system to measure an operation's effectiveness. The system must provide management with a few key indicators that reflect actual costs broken down into manageable units. Costs that manufacturing management cannot control (e.g., federal and state taxes, corporate charges) should be segregated (with care that such costs are, in fact, outside management's control).

Figure 12-6 represents an uncomplicated report format that reflects performance (costs are actual, adjusted for changes in inventory). Variances from standard costs (obtained from industrial engineering time studies or historical performance) are not included in this report because of controversy over the definition, validity, and accuracy of standards.

Standards are often used to conceal problems in order to show favorable operating variances. For example, if the standard cost of making a product is one dollar, but a high scrap rate increases actual cost to a dollar and a quarter, raising the standard cost will hide the scrap problem. Besides affecting operating efficiency, this inflation of standard cost tends to result in overpricing of the product and, thus, a noncompetitive product in the marketplace.

This article (*and text*) does not advocate renouncing the standard cost accounting system with its reporting of variances. The standard cost system can be used to value inventory; also, some variances (e.g., from standard pieces per hour or standard hours) afford a good measure of efficiency at an individual manufacturing work center because of the limited volume of items being measured there. Measurement of the efficiency of the total manufacturing operation, however, should use the actual costs of material consumed and labor

Sales	$1,000,000
Variable cost of sales	
Material	200,000
Tooling and supplies	200,000
Labor and fringe benefits	100,000
Total	500,000
Variable cost, percent of sales	50%
Factory overhead costs	
Controlled	100,000
Uncontrolled	100,000
Selling and administrative costs	100,000
Total	400,000
Income before tax	$ 100,000

Figure 12.6. Simplified Operating Income Report.

and overhead expended. Productivity measurements should be based on the total factors of production and not allow performance measures to overemphasize limited cost factors such as direct labor. (*As discussed in Chapter 4, the cost of quality (COQ) has become, and is likely to continue to be increasingly important as a factor in realistic cost measurement*).

The Product Costing System

Some proponents of Just-In-Time believe that the elimination of the cost accountant would enable successful speedy implementation of Just-In-Time techniques. Nonaccountants view current cost accounting systems as too restrictive and too complex to be of any value. However, because most manufacturing concerns will benefit from knowing their costs, the cost system should not be eliminated. It should be thoroughly reviewed just as manufacturing methods are reviewed before Just-In-Time implementation.

One method of cost system analysis is a thorough internal audit by a team of managers from various functional areas. The process reveals some interesting insights about how the system is perceived. The cost system is scrutinized and developed into a tool for the operating manager to use just as he or she uses a blueprint and production plan. The accountant should use knowledge of the current system to educate other team members and to prevent unfounded changes; at the same time, the accountant should be committed to endorsing constructive changes that may be suggested during the analysis.

As the manufacturing environment changes, both in terms of Just-In-Time and technology modernization, the external accountant should be one of the first to recognize that changes in the cost system may be needed. Recommendations presented to management, either by the internal or external accountant, should emphasize (1) the significant controllable cost factors and (2) simplification where possible.

Changover To Process Costing

The trend will be away from job costing toward process costing. Job costing refers to the costs accumulated for specific jobs or lots while process costing refers to costs accumulated by the process or operation. Process costing can be easily applied to all repetitive manufacturing conditions. It is much easier to maintain than job costing because the cost of processing the part or product stays the same until the process changes. Also, process costing requires fewer resources for data accumulation and processing and report generation. Thus, more resources will become available for more important functions when a process cost system is used.

Emphasis On Relevant Manufacturing Costs

Traditional methods of allocating costs to a specific product have emphasized direct labor costs. However, indirect costs (overhead) usually make up a much larger percentage of the total manufacturing cost than do direct labor costs. While modern machine tools make operations less labor intensive, they tend to require more supporting staff, such as technicians and programmers, to monitor their operation. The transition from direct labor to indirect labor renders obsolete those cost allocation systems that concentrate only on the direct labor cost of manufacturing. In many companies in which the processes are automated, direct labor is less than 10 percent of the total product cost; however, elaborate systems are still used to account for direct labor spending as if it were the major factor in manufacturing.

The absorption (or full) costing system also becomes difficult to defend because it allocates costs that have no relevance to the manufacturing of the product (e.g., property taxes, housekeeping costs). Just-In-Time production requires a costing system that identifies all relevant costs and incorporates them into the costing of parts. If additional overhead cost must be added for inventory valuation, this cost can be applied to the detail cost for financial statement presentation.

As the manufacturing processes change, the external accountant will be concerned that the costing of parts is based on sound judgement and logic. Accounting has often been, in the past, tied to the importance of consistency. However, consistent application of potentially outdated concepts, such as developing all overhead rates on direct labor, should be questioned.

Changeover To Hourly Pay Plans

In a Just-In-Time environment, pay plans will likely move away from piecework pay to some form of hourly pay. Piecework pay instills an attitude that quality is less important than volume. Because the objective of Just-In-Time is to produce precisely what is needed for a specific operation, quality parts are essential; the production of more parts than necessary represents a waste of labor and materials. Thus, a worker's pay cannot be based on the pieces produced. Management's attitude toward idle time must change because workers should no longer be producing excess inventories. In Just-In-Time production, the concepts of strict piecework standards, piecework checkers, and time clocks will become obsolete; in their place will be substituted effective managers and worker cooperation. As companies move toward that objective, a group incentive pay policy may be appropriate. Under this policy, workers would be paid according to the number of quality pieces that are produced. This form of payment will instill in the workers the importance of quality.

Determination Of Direct Labor Costs

A staffing table for an assembly line or work station should replace the traditional labor reports on which direct labor costing is based. The staffing would be determined by a study of the operations performed in a balanced line or work center and based on the volume of product required by the production plan. Some companies carry this costing concept even further; they consider direct labor a portion of the operating cost of the production line and allocate that portion of cost to the product that moves through the line.

De-Emphasis Of Machine Set-Up Costs

(*As we have seen in Chapter 5*), one of the objectives of Just-In-Time production is to reduce set-up times. In fact, set-up times are monitored by everyone in manufacturing as a factor to be reduced. Thus, when set-up times are reduced to minutes, the accounting system need not dwell on the cost of machine set-ups as it traditionally has done.

De-Emphasis Of Downtime Costs

Another important cost factor in the job shop environment is downtime. Each machine must run at maximum production in order to pay for itself, regardless of whether the parts produced are needed. In the Just-In-Time environment, the cost of downtime is not important if the machine is idle because the parts it produces no longer are needed; the motivation for minimizing downtime is the immediate need for the product rather than ensuring that the machine pays for itself.

The Inventory Valuation System

Traditionally, the accountant is responsible for reporting the value of inventory. However, the accountant does not order, schedule, or produce inventory, or perform the transactions that affect its balances. In effect, the accountant has no control over inventory quantities, but when there is an inventory loss, the accountant must provide explanations. Thus, the methods of inventory valuation change frequently.

The accounting system that buries problems in inventory will not support a Just-In-Time production system. The system that puts costs into inventory at actual rates and draws them out at standard rates will bury problems and produce surprises at the physical inventory. All variances must be forced out in the current accounting period by valuing the inventory from raw material to finished goods at standard cost. This valuation is done by multiplying both the beginning and ending inventory units by the same standard costs and accounting for the change in the monthly profit and loss statement.

The traditional preoccupation with the value of work-in-process inventory disappears as Just-In-Time techniques reduce the quantities of work-in-process. A stage-of-completion valuation is relatively unimportant when the amount of work-in-process inventory equates to 36 hours rather than three months. In fact, if work-in-process balances are small, it may be adequate to assume across-the-board that the total cost of any work-in-process inventory is one-half the finished inventory cost. (Of course, while work-in-process inventories are large, stage-of-completion valuation is, unfortunately, necessary for the accurate valuation of work-in-process). (*These improvements will take time, and therfore require stage-of-completion accounting support during the transition*).

Various accounting benefits will result from the reduced inventories that accompany Just-In-Time implementation. The number of transactions decreases as the number of storerooms is reduced and the number of work-in-process units declines. Also, dimished are the needs for obsolescence reserves and for reserves to compensate for losses at the physical inventory. Furthermore, in traditional accounting systems, attempts to achieve inventory accuracy can be costly and time-consuming; reducing these expenditures is another benefit–low inventories allow easy verification of inventory accuracy.

As Just-In-Time production is achieved, the traditional inventory valuation system should be replaced by a system that motivates the supervisor to verify the work-in-process inventory daily within his or her production area. Work-in-process inventory should be small enough to permit easy determination of quantities. Daiy assessment of work-in-process inventory simplified, for example, by the use of containers that hold standard quantities of parts will allow easy identification problem areas if quantities suddenly increase. Confidence in the accuracy of the work-in-process figures will result in immediate attention to problems.

Other Accounting Areas Affected By Just-In-Time

Just-In-Time implementation will affect a variety of other accounting practices; some of these are discussed in the following sections.

Investment Justification

When an investment opportunity is evaluated, the justification process typically includes all costs,including indirect ones, as they relate to the specific investment. Many desirable investments will appear unattractive unless an analysis of the total potential savings (including the costs of inventory and quality) is also made. Quality is crucial to successful Just-In-Time production. If an investment can improve quality, but those improvements are difficult to quantify, then a company committed to Just-In-Time must be prepared to examine its return-on-investment criteria. For example, the existing approval levels required for an investment may prevent lower level managers from attempting new processes that could benefit the business. Relaxation of approval levels to simplify the justification process may be necessary. Representatives from all manufacturing disciplines should be involved in evaluating the total relevant costs and benefits in the justification of investments as well as in make-versus-buy decisions.

Accountants, both internal and external, need to understand the important success factors within the business and not just account for or audit individual transactions. A critical look at the capital investment strategy and mechanics used by a company provides a good indicator of the adequacy and pervasiveness of the control over capital investment. The indirect costs of the past must be better managed to realize the benefits that should be obtained from such investment. It is no longer adequate to assume, for example, that labor and overhead relationships will be static when estimating project costs and benefits.

Vendor and Customer Relationships

As the organization moves closer to Just-In-Time deliveries and shipments, relationships with vendors and customers require reassessment. The practices of the purchasing and accounts payable groups will likely change.

Competitive Bidding

(*As we have discussed in Chapter 7*), vendors of purchased components will be selected not on the basis of their competitive bids but according to criteria such as the total cost of quality, geographic location, delivery capabilities, and financial stability. In effect, vendors will become extensions of the in-house manufacturing effort. Prices will be negotiated when the vendor signs on as a supplier and will be updated periodically. Accountants will be responsible for verifying the financial stability of the vendors in much the same way the quality and engineering groups verify a vendor's capability to produce parts to the required specification.

Invoices and Purchase Orders

Invoices and purchase orders for parts received and supplied on an ongoing basis will become obsolete. Vendors will ship and be paid according to a production schedule supplied to them by the manufacturer. Likewise, customers will supply their production schedules and will pay according to their receipt or consumption of the goods. The previously negotiated prices will be input into a computerized payables program.

The system will eliminate the need for the volume of paperwork traditionally required to order the same parts from vendors and to ship the same products to customers week after week. However, success of the new system requires defect-free products as well as a thorough understanding among accountants at each end of the transaction (manufacturer and vendor, manufacturer and customer).

No area will have a more pervasive impact on the external accountant than the interfaces between the company and its vendors and customers. Determining inventory ownership becomes more complicated as increased vendor assist operations are performed. As paperwork is eliminated and inventory movements are pulled by a recurring production schedule, controls must be established to test, for example, vendor payments with defect and product shipment data.

Systems Design

The accountant is only as good as his or her tools. The accountant must be creative enough to design the systems, report formats, and computer screens that will support the programs on the shop floor and reduce the often columinous paperwork required in the office.

One suggestion for a new tool is an event-activated system for recording transactions–that is, a system in which a specific action activates all the related transactions needed to record the event. For example, a receipt of goods from a vendor can activate the following transactions:

- Input into inventory
- Delete from outstanding purchase order (possibly a purchase order for a vendor's capacity; specific order quantities are determined by production releases)
- Trigger payment to vendor by wire transfer

- Delete open material commitment
- Record transaction in cost-of-goods-purchased account
- Calculate material spending variance
- Record receipt data and quantity in vendor history file (to enable later assessment of vendor performance)

This same kind of system can be developed to activate transactions for sale of a product, completion of a manufacturing operation, receipt of an order, or change of a bill of material. The computer is essential in all of these actions; to ensure confidence in the computer's performance, the accountant must set various control limits and use exception reports.

The external accountant will support the integration obtained from such an even-activated system. The ability to perform advanced computer-assisted auditing is greatly enhanced by an integrated system. As systems have become so complex at certain companies, it has become exceedingly difficult to develop a written integration plan. Such integration planning is required to simplify today's system.

Concluding Comments: Clarifying the Accountant's Role

In a Just-In-Time production system, accountants must concentrate their reporting activities on providing information that helps managers focus on problem areas. An accountant who assumes the role of overzealous protector of the company's assets and concentrates on outdated policies will discourage well-intentioned managers.

The enlightened accountant must be part of the solution to problems that surface during the implementation of Just-In-Time production, not an obstacle to the adoption of the Just-In-Time philosophy. Thus, the accountant is no longer the scorekeeper. Failure of the accountant to accept the new role will represent a major obstacle to Just-In-Time implementation.

Failure of the external accountant to encourage change also will slow down the implementation of Just-In-Time concepts. It is the responsibility of all accountants to understand the implications of Just-In-Time.[1]

Employee Involvement

It is impossible to accurately predict the potential value of Employee Involvement program. It is useful, however, to review the progress of any existing programs such as employee suggestion and/or work simplification programs. The specific data needed includes the number of suggestions submitted per employee, the percentage of suggestions adopted, and the savings generated from the program as well as average dollar savings per suggestions submitted/adopted. It is not uncommon to find EI programs, because of their structure, focus, and training aspects are more than twice as successful in terms of savings dollars generated over a one-year period. EI programs typically can be expected to cost about 5 to 10% of employee time, but are self-funding in less than one year. However, because each organization is so different, this information can only really be used for benchmarking, not for accurately estimating potential savings.

Summary

Opportunity assessment, while more extensive in its initial application, needs to be performed periodically to support continuous improvement activities. To support the development of the implementation plan, the assessment team should carefully document and retain all findings and supporting data.

It should then be possible to dollarize potential savings, prioritize them by area, apply general implementation cost estimates, and construct a cost/benefit analysis for management based upon these findings. It is generally advisable to break potential savings into "hard" and "soft" dollars, based on traditional management criteria. At least preliminary plans should be drawn up for how reduced manhours will be reapplied, and/or redeployed as well.

References

1. William G. Holbrook, Robert Eiler, "Accounting Changes Required for Just-In-Time Productions," Synergy '86, 1986 Conference Papers, (Dearborn, MI: Society of Manufacturing Engineers).

PILOT PLANNING AND DESIGN 13

The phase described here as Pilot Planning and Design involves selection of the pilot project(s), selection and organization of the pilot project team(s), development of a macro-level overall facility layout, development of a detailed layout for the pilot area(s), identification of both general and specific pilot project performance measures, development of a general (macro-level) implementation plan, and development of a specific implementation plan for the pilot area(s).

Pilot planning and design takes the organization through the fundamental planning process that will ultimately produce the framework of the implementation. Selecting the appropriate pilot(s) and planning their implementation is analogous to selecting a child's first piano instructor. Not only can a poor choice impede the progress of the child, but it might instill bad practices and eventually destroy all enthusiasm for a lifetime. A sound choice, accompanied by appropriate encouragement and early success, can foster a lifelong practice of continual improvement leading to world-class stature.

Pilot Project Selection

There seem to be two basic approaches to pilot project scope selection: 1) The open-ended cell approach, and 2) The complete line approach. Experience suggests that the complete line approach is preferable.

To visualize the advantages of incorporating changes over the entire production routing of a product line rather than individual operations in departments, consider *Figure 13-1*.

The product is produced in a routing that has operations in each department. The routing is A3, B3, C3, D3, E3, ship. Each department has setup times of 50% of the total production hours. There are two approaches to waste elimination

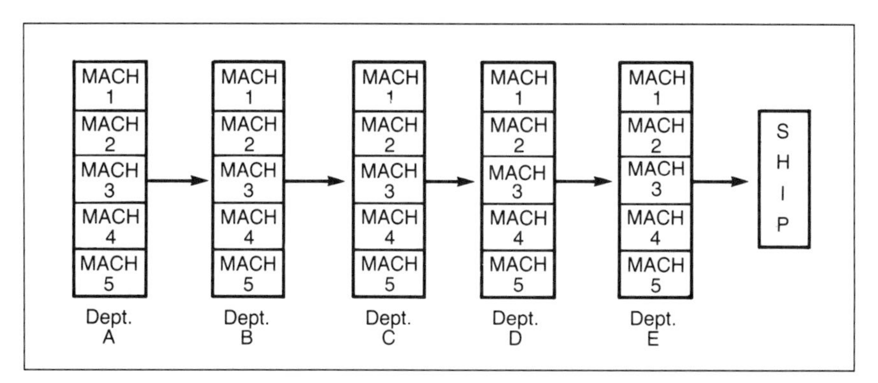

Figure 13-1. A sample production routing of a product line through individual departments.

in the form of setups: One is to reduce all setup times in one department, and the other is to attack only those setups associated with the product line involved, but in every department.

In a scenario of the first approach, the setup times for an entire department (say Department C) are reduced by 50%. In this case, then, since 50% of 50% of the time in this department was eliminated, 25% of Department C throughput time is removed. Since Department C contains 20% of the total production routing throughput time, the actual throughput time is reduced by 25% of 20%—or 5%.

In a scenario of the second approach, only the operations associated with the production of this product are targeted for setup reduction, but in every department. Again, setups are reduced by 50%. Since setup times represent 50% of total throughput time, the time saved is 50% of 50%,—or 25% of actual throughput.

The difference, then, is clear. In terms of the benefit, it is the difference between a 5% improvement in all product lines that include Department C in their routing and 25% throughput time improvement in the target product line. (Similar relationships exist in terms of inventory reduction and overall product cost). The choice is between "islands" or small units of efficiency, and substantial overall productivity gains that are visible to the customer.

Assuming, then, that the entire flow involved in a specific product line will be addressed, which product line do you choose as a pilot? There are a number of things to be considered. First of all, the product line involved should be a "significant-but-not-major" part of the total production volume. Typically, this means somewhere between 5 and 15%. That amount will provide an adequately credible example, without involving a massive infusion of effort and resources. Secondly, the pilot product line should be fairly distinct and easy to isolate. This will minimize disruption to total factory operations. Thirdly, the pilot product line selected should be fairly representative of the remaining product lines whenever possible. In this manner, the problems that arise in the pilot project will provide a sound training ground for remaining factory conversion, and they will speed up the balance of implementation activities.

In those cases where widely varied product lines and processes are involved, it may be desirable to utilize more than one pilot line. However, it is important to bear in mind that multiple pilot lines will involve multiple resource requirements, training, etc. In addition, the administration and control issues become larger.

Pilot Project Team Selection and Organization

The selection of pilot project team members should be based on the candidates' enthusiasm, understanding, and technical ability. In other words, the pilot project team members should be selected for their ability to make significant contributions, as opposed to their mere availability. The team should include personnel from these disciplines:

- Materials Management (purchasing, production control, and/or production planning).

- Industrial and/or Manufacturing Engineering.
- Design Engineering.
- Quality Assurance.
- Manufacturing Line Supervision.
- Organized Labor (where appropriate).

In addition to the core team members, expertise may be required periodically from outside consultants and/or other specialists within the firm. Also, specific areas such as setup reduction may utilize subgroups (teams), including a ''ringer'' as discussed earlier.

The original pilot project team should remain in place throughout pilot implementation. At that point, it may useful to distribute successful pilot team members to subsequent remaining factory conversion activities. The ideal pilot project team size seems to be about seven people. Anywhere from five to ten is acceptable.

The team should be assigned on a full-time basis, with weekly status reports and typical project management activities. Administration and day-to-day coordination activities should be handled by the team leader. The leader can be any team member, and he/she should be designated by the steering committee when the pilot project team is first organized. The pilot project team usually reports to the JIT program steering committee every two-to-four weeks regarding their progress.

The objectives of the pilot project team are to generate a macro-level overall facility layout, a detailed pilot line design, and a detailed implementation plan for the pilot line.

Macro-Level Overall Facility Layout

The macro-level overall facility layout, sometimes referred to as a conceptual design, is a ''vision'' of future operations. It should incorporate the concepts of quality, balanced work loading, reduced setups, cellular manufacturing, etc. in it's construction. Based on the utilization of these concepts, and maximizing the use of focus factory techniques, it should be possible to lay out overall factory flow and space requirements. It will be important to remember that you want to include only value-adding activities and eliminate (or at least minimize) WIP inventory stores/staging/queue areas. Rough square footage requirements based on machine dimensions, general aisle requirements, and space for tooling and equipment may be used to rough-in work center floor space in blocks, once general product flow has been determined. Remember, too, that while purchased part/raw material receipt should be considered, it may be altered so that receipt occurs at or near the point of use.

The macro-level design, then, will culminate with a block diagram layout of the new factory envisioned. Depending on the requirements of senior management, it may also be necessary at this point to perform a cost/benefit analysis of the envisioned overall layout. Experience, however, has shown that this is never completely accurate. There is often a large number of unresolved issues and unknown data at this stage. A pilot implementation provides concrete evidence of likely benefits to be gained by conversion of the remaining factory. It is usually the larger, more bureaucratic organizations that require this kind of analysis and formal reporting. When it is required, it makes a great deal of sense to have it prepared with the assistance of experienced, independent consultants. Unfortunately, in order to perform a thorough cost/benefit analysis and develop pro forma operating statements, a great deal of time and resource expenditure is required. Some recognized consultants expend as much client time and resources developing this documentation as was originally required to implement pilot activities.

Detailed Layout for the Pilot Projects

When the overall floor layout is complete, it should be possible to take the next step, and construct a detailed model or layout of the pilot project operations. All of the information gathered during the opportunity assessment should be utilized in the development of this design. Critical considerations include:

- Current and projected product line demand.
- Current and expected equipment availability, capability, capacity, and utilization.
- Opportunity for utilization of cellular manufacturing techniques.
- Expected quality levels, and the need for test equipment.
- Specific work center cycle times, setup/change over times, and parts produced on each.
- Operator cross-training levels, current and required.
- Manpower levels, current and required.
- Anticipated frequency and quantity of purchased part/raw material receipts.
- Current and anticipated levels of part standardization. (The reduction of the number of components in a subassembly can result in proportionate reductions in storage space required, and this can have a significant impact on the overall floor layout).

The output of the specific pilot project layout should include a detailed drawing of the area involved, including product flows, machines used, and all anticipated storage areas for materials, tools, and equipment. In addition, balanced work loads and the number of kanbans between operations must be calculated. Various manning levels should be figured and documented to meet demand fluctuations, and cross-training requirements for specific cells and work centers should be identified. Items that should not be overlooked are material handling devices. Included in this category are flow racks, conveyors (both roller and powered types), cranes, forklifts, and electromagnets. While these devices are common enough, they all have strengths and weaknesses, as well as significant limitations (including cost).

Another more obvious consideration is often referred to as "monuments." Monuments are pieces of equipment that pose serious problems in terms of being relocated because they are mounted on deep foundations and/or attached to special hydraulic, pneumatic, or coolant devices. Often it isn't possible to begin with a clean slate in terms of floor layout, and in cases where machines must be moved, the situation has been rightly compared to a game of checkers. You must have an open space into which you can move your players. In addition, correctly timing and sequencing equipment movement in order to minimize disruption to the operations are critical to successful realignment of the facility.

Performance Measures

Performance measures associated with the Just-In-Time program should be developed to support the overall company goals and objectives, as defined in the company's strategic business plan. Typical company business plans include improved customer service levels, increased market share, reduced cost-of-goods-sold, and improved product quality levels. These are broad categories, within which a number of more specific objectives will be needed. The more specific objectives may be constructed from the faces of the JIT "emerald" model shown in Chapter 3, *Figure 3-1*. Among the performance measures that might be generated from this model are:

1. Quality.
   - Scrap cost reduction.
   - Debit dollar reduction.
   - Reductions in customer complaints per time period.
   - Reductions in scrap and maintenance as a percent of sales.
   - Rework cost reductions.
   - Reductions in QA budget as a percent of sales.
   - Reductions in reported cost of quality.
2. Simplified, Synchronous Production.
   - Increases in percentage of parts run to schedule.
   - Reductions in setup time divided by number of setups.
   - Increases in "okay" first pieces divided by number of setup attempts.
   - Reductions in setup time per pieces produced.
   - Increases in the number of balanced work centers divided by total number of work centers.
   - Reductions in WIP inventory dollars and volume.
   - Reductions in finished goods inventory dollars and volume.
   - Reductions in the number of machines with debris visible around machine divided by total number of machines.
   - Increases in the number of machines on systematic preventive maintenance divided by total number of machines.
3. Process-Oriented Flow
   - Increases in the percent of activities that add value.
   - Increases in the number of finished pieces produced divided by total man hours.
   - Reductions in overall travel distance.
   - Reductions in total WIP and finished goods inventory dollars and quantity.
   - Reductions in the average manufacturing throughput time by product line.
   - Reductions in the number of reportable injuries divided by total man hours worked.

4. Advanced Procurement Technology
    - Reductions in raw material dollar and quantity levels.
    - Reductions in quantity of defective raw material divided by total quantity of raw material received.
    - Reductions in the number of suppliers per commodity.
    - Increases in the number of suppliers involved in the supplier certification program divided by total number of suppliers.
    - Increases in the number of suppliers involved in joint cost reduction and/or quality improvement programs divided by total number of suppliers.
    - Reductions in the average procurement lead time by commodity.
    - Increases in the number of suppliers within a 250-mile radius divided by the total number of suppliers.
5. Improved Design Methods
    - Reductions in the average processing times for engineering change orders.
    - Reductions in annual volumes of engineering change orders.
    - Reductions in the average number of parts contained in engineering change orders.
    - Reductions in the average number of operations involved in new designs.
    - Reductions in the average number of tools required to produce new designs.
    - Increases in the percentage of employees involved in Value Analysis/Value Engineering/Current Product Review programs.
    - Reductions in overall product cost.
6. Enhanced Support Functions
    - Reductions in reporting points for production in the factory.
    - Reductions in indirect labor/overhead assigned to production scheduling and/or expediting activities.
    - Reductions in volume of labor-reporting transactions.
    - Increases in the number of direct-labor employees divided by total number of employees.
    - Reductions in the average distribution throughput time (shipping dock to customer hands) by product line, by distribution channel, and by geographic area.
    - Reductions in the average distribution cost by product line, by distribution channel, and by geographic area.
7. Employee Involvement
    - Increases in the number of employee suggestions divided by total number of employees.
    - Increases in the savings generated by EI teams divided by the number of EI hours expended.
    - Increases in the percentage of employee suggestions implemented

This list of performance measures should not be considered comprehensive, nor is every performance measure appropriate for every organization. In fact, there are more than 200 such measures on file. Traditional measures such as inventory turns are still appropriate in many cases, but they are redundant for the purposes of this text.

When establishing performance measures, it is critical to identify the source of future measurement data. For example, what specific report, log, or database will be queried to obtain the particular measurement information over time? Consistency and appropriateness (relevance/accuracy) of the data are crucial.

General Implementation Plan

The general implementation plan is a macro-level plan that sequences the major activities involved in overall factory conversion and schedules them in a manner that facilitates benchmarking of program progress.

The general implementation plan usually begins with a view of the envisioned operation such as the block diagram layout described in the "Macro-Level Overall Facility" section mentioned earlier in this chapter. As also discussed, it is necessary to perform any relocation of equipment in a manner that minimizes disruption to ongoing production operations. A "place to move the checkers" is needed to start the process and to move equipment each time so that space is left for the next move. (Like most things, this is easy to say but difficult to do. Utilities, foundations, hydraulic lines, air lines, coolant lines, and ventilation equipment must be considered in the timing of these activities). Sequencing these major activities should allow for the development of a Gantt chart, based on the primary tasks within each major activity. Consider the layout shown in *Figure 13-2*:

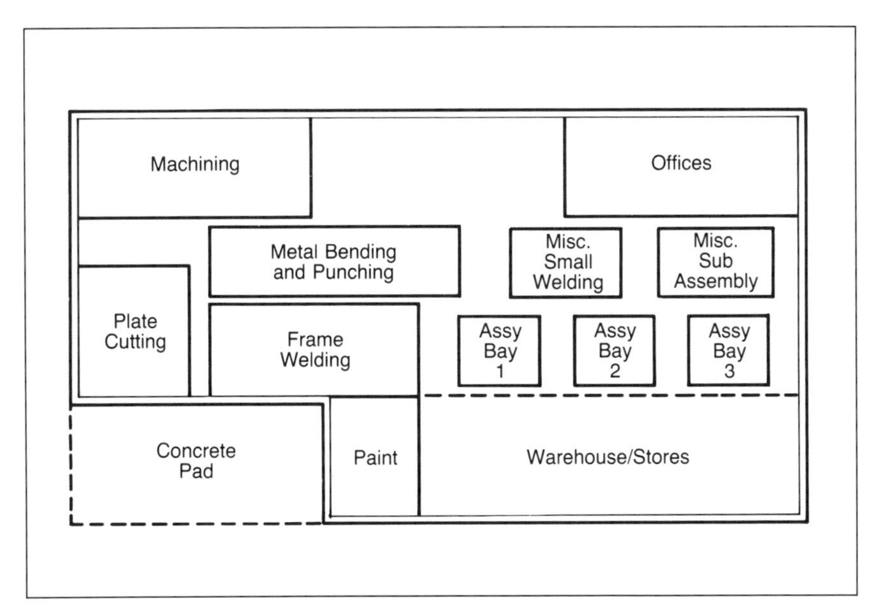

Figure 13-2. An existing plant layout.

In this example the major activities involved in the general implementation plan might be:

1. Enclosure of the concrete pad to increase available floor space.
2. Movement of most frame welding to a newly enclosed area.
3. Construction of (welding and subassembly) pilot project, including all operations starting with cut plate and ending with delivery of major subassemblies to final assembly.
4. Conversion of assembly bays into parallel final assembly lines, with worker movability between lines.
5. Construction of two other welding and subassembly projects similar to the pilot project, and the freeing up of the miscellaneous small welding area.
6. Conversion of bays on each side of the parallel assembly lines into subassembly stations for major purchased components, and the facilitating of direct feeds to line.
7. Purchase and installation of flow racks to support subassembly operations.
8. Relocation of the paint area.

The layout for future operations, would then appear as shown in *Figure 13-3*:

The incorporation of cellular manufacturing techniques, pull systems, and balanced (uniform) work loading will allow for the compression of square footage requirements, and improved purchasing techniques should reduce the amount of incoming material space requirements. Warehousing and store areas can be utilized for service parts, and so forth.

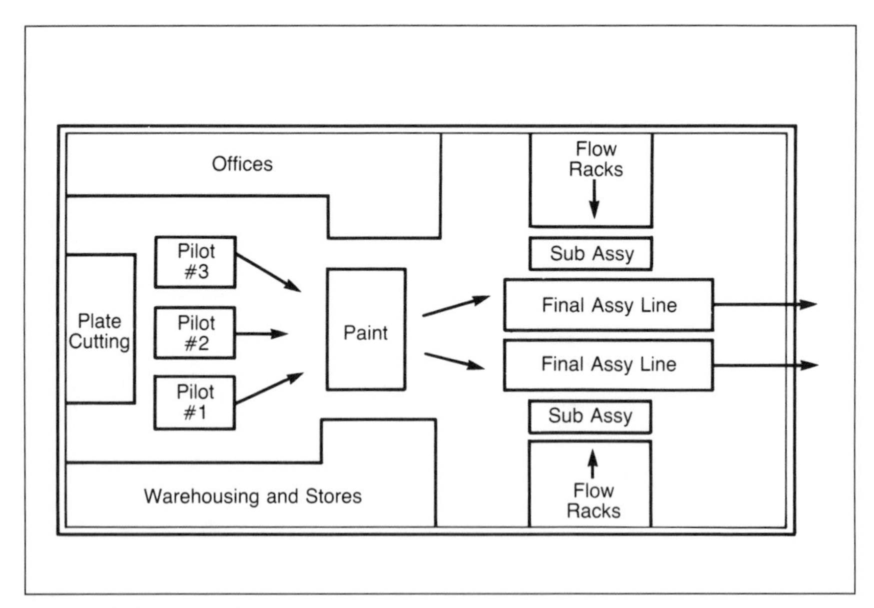

Figure 13-3. A new plant layout.

Next, each major step will need to be broken down into it's primary tasks, with time estimates allocated to each task. The time estimates may then be "rolled up" to construct a Gantt chart of overall implementation activities, as shown in *Figure 13-4*:

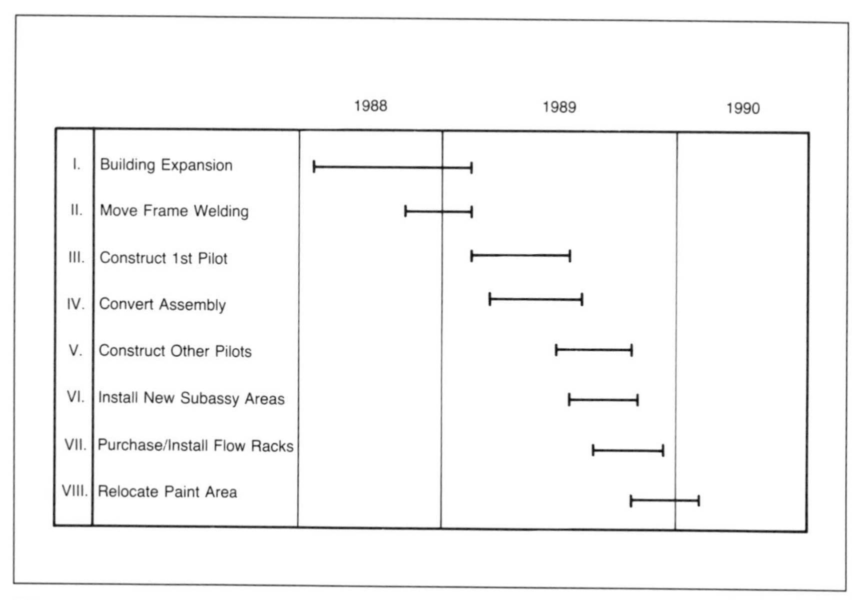

Figure 13-4. A Gantt chart of a Macro-Level Implementation Plan.

The kind of primary tasks to be considered in a major activity such as Building and Expansion would include steps like these:

1. Initiate and complete architectural design for building expansion. (Identify relocation of materials currently stored on pad).
2. Review architectural plan to ensure proper door locations and ventilation considerations for paint and welding.
3. Identify any dock requirements.
4. Identify any requirements for new outside concrete pad.
5. Identify inside (existing) wall change requirements.
6. Initiate and complete building expansion construction work.

It is important to note that all steps of the implementation plan will not fit neatly within logical Gantt chart time frame groupings. For example, in the case listed above, it might well be necessary to order new paint equipment and/or material handling devices during the same time period that building expansion occurs. The point here is that macro-level benchmarks should be established at this stage for overall program implementation. These benchmarks should be determined based on an intelligent assessment of the primary tasks involved in each major activity, and on the time required to complete them. As the pilot project is completed, additional levels of detail will become known in terms of equipment requirements, manpower requirements, etc., which can then be built into the overall implementation plan.

Specific Pilot Implementation Plan

Depending on the areas that are chosen to become involved in the pilot implementation, the project plan can span time frames ranging from three to eighteen months. One of the real secrets to minimizing implementation time is sound up-front planning and design.

In the specific (or detailed) pilot implementation plan, both the actual floor layout and the associated steps to achieve the envisioned flow will need to be defined more carefully. With regard to floor layouts—whether the setting is the manufacturing floor or the office—the design should usually be done using a 1/4 inch = 1 foot scale, with all work centers and material (or paperwork) flows noted. The dimensions, reach, and capabilities of all material handling devices should be considered, along with required space for all tools, gages, containers, and WIP material. Adequate work area must be allowed, along with sufficient clearance for WIP movement and machine access for preventive maintenance.

The design should maximize the utilization of cellular techniques and minimize the inclusion of nonvalue-adding activities/material handling/travel distance/throughput time. Flexible manning levels should be accommodated, along with movement of operators between work centers.

In terms of the implementation plan itself, it is important to remember that in conjunction with the conversion of the physical process, supporting systems must be developed and implemented as well. This means that significant changes will occur in quality control, production control, purchasing, design engineering, and accounting areas. In addition, distribution functions may be affected as well as industrial and manufacturing engineering areas. It is useful when constructing the pilot implementation plan to consider and address each face of the JIT "emerald" model. While there is no set sequence of major activities that applies in every case, *Figure 13-5* depicts a popular general flow.

It should not be concluded, based on this example, that any of these items other than organization really have a completion point; only initial implementation can be completed. All other activities are subject to continuous improvement. Implementation of the Advanced Procurement Technology (APT) can take up to two years, depending on contract renegotiation, etc., hence the arrowhead at the right hand side of the APT time line.

As a rule of thumb, the detailed pilot project implementation plan should be two levels deeper than the Gantt chart shown in *Figure 13-5*. (Think of the levels in a typical bill of material). For example, the major activities listed under Improved Design Methods (IDM) should include: Streamlining ECN Processing, Cost Reduction Programs (VA/VE/CPR), Design for Quality, and Design for Manufacture. Under each of these headings, there should be sequential listing of primary tasks. For example, under the heading of Streamlining ECN Processing, the primary task listing should include such items as:

1. Organizing ECN task group.

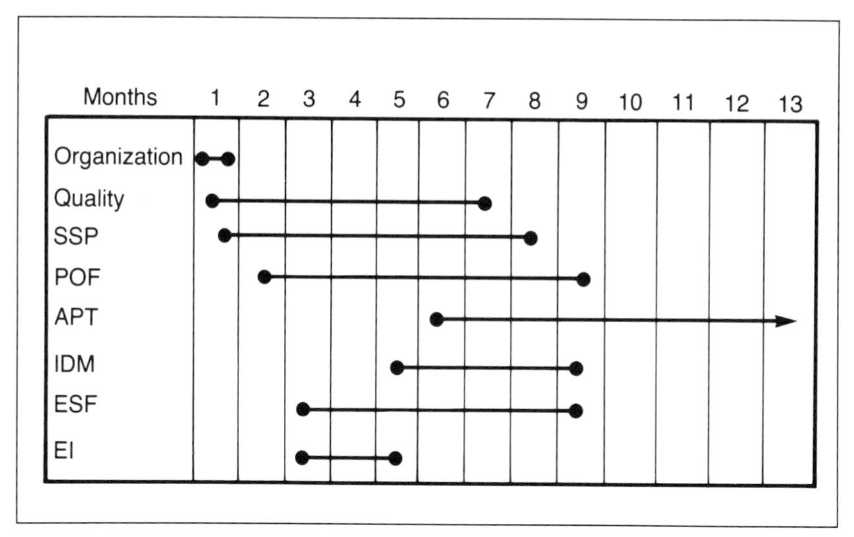

Figure 13-5. A chart of a General Implementation Sequence.

2. Performing any required assessment training.
3. Documenting existing ECN processing (throughput times, specific activities involved, responsibilities, volumes, etc.).
4. Identifying any duplication of effort, unnecessary activities, opportunities for error, timing problems, etc., which result in nonvalue-adding activity.
5. Performing any required problem-solving training.
6. Conducting brainstorming/problem-solving/cost-benefit sessions to identify and prioritize waste elimination opportunities.
7. Documenting the best recommendations, and reviewing them with affected personnel. Making any required adjustments to recommendations, based on feedback.
8. Presenting recommendations for streamlining to JIT steering committee and to any other appropriate members of management.
9. Making any required adjustments to recommendations, based on feedback from management and the steering committee.
10. Documenting final recommendations, and gathering any remaining details needed for implementation. Writing procedures for new methods.
11. Training affected personnel in new methods.
12. Establishing appropriate auditing methods and performance measures to ensure compliance with procedures over time.
13. Implementing new methods.
14. Implementing audit and reporting processes to monitor compliance levels.
15. Implementing ECN processing performance measurement and reporting results.
16. Dissolving ECN task group.

When the primary task listing has been completed, the implementation plan can be constructed. One implementation plan format used most successfully includes the task listing and the following additional information for each task:

- *Responsibility*–The name of the individual who bares ultimate responsibility for the completion of this specific task.
- *Start Date*–The date on which the specific task should be undertaken.
- *Original Completion Date*–The date by which the specific task was originally scheduled to be completed.
- *Current Completion Date*–The date by which the specific task is currently scheduled to be completed.
- *Deliverables*–The circumstances, results, or evidence which will confirm that the specific task is complete.

The format of the implementation plan constructed from these elements will look something like that shown in *Figure 13-6*:

STREAMLINING ECN PROCESS					
Activity	Resp.	Original Start	Comp.	Current Comp.	Deliverables
1. Organize ECN Task Group	J. Smith	3/5	3/10	3/10	Operating task group
2. Perform any required assessment training	B.Duncan	3/10	3/15	3/15	Completed task group training
3. Document existing ECN process	G. Jones	3/15	4/30	4/30	Completed flowchart with throughput times, activities, and responsibilities.

Figure 13-6. A Sample Implementation Plan Excerpt.

When the entire pilot project implementation plan is laid out, it should be loaded onto a good project management software package. Project management software will assist in maintaining visibility of progress to plan and in ensuring the integrity of dependent date relationships as completion dates are modified.

PILOT IMPLEMENTATION 14

The implementation of Pilot Project Operations involves assignment of personnel to the pilot, initiation of pilot line plan activities, implementation of support systems, monitoring and reporting of pilot project progress, problem identification and resolution, and preparation of leaders for technology transfer to support the remaining facility implementation.

The success or failure of the pilot project is a pivotal point. Failed pilots usually spell disaster for the program, since they expend management commitment and enthusiasm, and employee buy-in without significant improvement in return. In addition, the pilot project establishes the foundation of technology and training which will be "leveraged" into other projects, until the remaining facility has been converted to Just-In-Time operations.

Assignment of Personnel

The importance of the pilot means a good deal of thought should be given to the selection of the people involved. In every organization, there is a percentage of people who push themselves to improve. These people take night courses, work toward certifications, and work beyond regular hours. They are often informal leaders, people who are respected by their peers, and whose opinion is sought by others.

Some are technically skilled, while others are skilled in the political workings of the organization. These are the kinds of individuals who make the difference in pilot project operations. They are tenacious enough to battle through technical and political difficulties, and innovative enough to generate creative approaches in problem solving and day-to-day operations.

There are, however, dangers in "loading" any group with too many of these types of people. Often, they have very strong personalities and egos. They have a tendency to be mavericks, and go off in their own directions. They are difficult to manage. On every pilot project team, there should be a few of these people (from management, and/or the ranks of machine operators and administrative staff). To balance these people, each team also needs a contingent of experienced, methodical individuals, respected for their seniority, stability, and prior contributions. These people are "team players," trusted by management and with a history of high quality work.

Finally, each team requires leadership. The team leader should be aggressive, and have a track record of successful project management. He must believe in the need for Just-In-Time manufacturing practices, be politically "connected" enough to get what he needs from management, and be strong enough to control the mavericks, motivate the stable members, and drive the implementation process without stifling the team. This individual must champion the implementation program, and accept without qualification the responsibility for its success.

The implementation team will involve individuals from the support areas such as purchasing, production control, design engineering, industrial engineering, accounting, distribution, and, when appropriate, MIS, as well as the machine operators involved. The entire team should meet periodically to review implementation progress and any problems which may be surfacing. Subsets of the team will be meeting periodically to perform their individual tasks throughout the course of implementation.

Initiation of Pilot Line Activities

Active pilot plan implementation activities should be initiated by the team leader at a kickoff meeting. The kickoff ensures all team members know who each other are, establishes visibility that the program is underway in the rest of the organization, and creates ground rules. Team members should be encouraged to introduce themselves, telephone numbers at both work and home should be exchanged, and primary responsibilities should be described. The kickoff meeting can go a long way toward an important bonding process for the team. In addition, it should include an overall picture of the time frame of the pilot implementation, and how it fits into the general implementation plan. This will reinforce a sense of purpose and program continuity, and provide a roadmap for the future. Finally, the kickoff meeting should express genuine gratitude and respect to the people involved in the pilot implementation. They will be challenged to constantly improve not only the way they do their jobs but how they *think* about their work and the business.

Implementation of Support Systems

The support systems constructed to facilitate Just-In-Time operations will need to be developed, implemented, and operational for pilot project areas at the same time traditional manufacturing and support systems are running in the rest of the facility. The primary disadvantage of this situation is that resources will be required to perform the implementation and maintenance in addition to operating the standard systems. This frequently occurs when new systems are installed. Few systems run flawlessly from the start. There is usually a period of working out bugs and verifying output integrity, while the old system continues to function. This is known as "running in parallel," or a "parallel system" environment. The advantage of this approach is that it allows problems to be identified and resolved in a relatively controlled setting. By the time the support systems are responsible for the operation of the entire factory, they have been tested and improved gradually through multiple pilot projects.

In production planning, it will be necessary to determine appropriate lot sizes for pilot operations, and scrap develop allowance generation practices. Since the theoretical ideal for lot sizes is one piece, it will be important to track setup times and cycle times, reducing lot sizes slowly as setup time reductions, demand stability, and work load balance are achieved. The most significant impact in terms of lot sizing is that it becomes a function no longer performed by production planners in a front office, but migrates to a shop floor control function.

Typically, the lot size is a function of cycle time and end item demand. These two factors are variables controlled on the shop floor in a JIT environment, and the determination of what lot sizes will be used should occur on the floor. Lot sizes may be varied within cells, for example, just as cell manning levels are varied. Depending on how the supervising organization is structured, lot sizing may be determined by a "cell leader," "focus factory manager," or the department foreman. Ultimately, lot sizing for components, subassemblies and assemblies must meet end item demand. The general rule is that a constant effort should be made to reduce lot sizes over time as better balance is achieved, setup times are reduced, and quality levels are improved. The calculation and generation of scrap allowances should also be minimized as quality levels are improved. Some organizations simply generate a predetermined level of WIP and/or finished goods and lock it up. In the event of a major quality problem, the material is withdrawn and used. This may be viewed as "begging the issue," but it has allowed a number of companies to overcome initial jitters, and move on to successful implementation. In most of these cases, the "kitty" or materials disappears over 4 to 6 months.

Another aspect of production planning diminshed by Just-In-Time manufacturing is generation, maintenance, and closing of production orders. Individual production orders for manufactured components, subassemblies, and assemblies often become unnecessary as a pull system is used to link production. Actual manufacture is sequenced via kanban from end product assembly back through the disbursal of raw material. A final assembly schedule is usually generated from the master production schedule, with adjustments made beyond a predetermined "frozen" horizon, based on availability of tools, equipment, materials, manpower and demand. When the final assembly schedule is produced, all of the component production may be reported automatically. Inventory may be "backflushed" all of the way back to raw materials as well. When systemic, individual component and/or subassembly orders are used, they may be closed out automatically as well based on production of end products. Many companies have taken this route, because they perceive it as easier than modifying systems to not produce manufacturing orders for components.

What is done with production planners when most of their primary responsibilities go away? Generally, their work has forced them to develop contacts throughout the factory, and they are used in looking at the "bigger picture" of factory operations. They may be considered for mid-level management training programs, or reassignment and training in a related area like purchasing.

Production control is similar to prodction planning because its primary functions are minimized in a Just-In-Time environment. Likewise, they do not disappear overnight. However, as pull systems become the sequencers and expediters of production, and as individual machines are no longer scheduled independently, the need for individuals dedicated to these tasks lessens. "Hot list" meetings and parts chasing should be unnecessary. The first-in, first-out (FIFO) nature of pull-driven production should make these activities obsolete.

Production schedulers have long been "trainees" for line supervision positions. Their view of operations is often focused at the departmental level.

Displaced schedulers and expediters often end up supervising manufacturing operations at the department or cell level.

Accounting systems changes required to support JIT operations have been a topic of debate. In general, cost accounting is streamlined. Fewer labor reporting points and transactions are used because there is less numerical control required. In an environment where production operations are physically close, work in process levels are minimal, and material does not move to or from stores between the first and final production operations, there is little need for issues, receipts, and transfers. When little WIP exists, there is no point in reporting individual production operations, since the WIP will be finished goods before it can be accumulated and valued for constructive management use. Often, the traditional categories of inventory (raw, WIP, and finished) are reduced to two: RIP (raw and in process) and finished. Direct labor, rather than being reported on an individual part number and pieces per hour basis, may be ''backflushed'' like inventory, if approximate values for preventive maintenance, process control charging, and housekeeping activities are accepted. Each organization must develop a system which best meets its own needs. The implementation of the system changes in a pilot setting should allow most bugs to be worked out in a controlled setting, so the methods may be proven and accepted by management before being transferred to the balance of the operation.

Accounts payable and accounts receivable improvements, including the use of EDI and general streamlining of paper work processing, may be implemented at any point. However, initiation of improvements in these areas provides a ''sense of belonging'' and being part of the improvement team as pilot operations get underway in the factory. It is good timing from the standpoint of motivation and morale. General streamlining of the paper flow should be attacked first. Thorough charting by an objective party usually uncovers many areas where activities are redundant or unnecessary. In addition, ''holes'' are often discovered where information is lost altogether.

Problems in the paper work process, like problems in production, should be systematically attacked through problem solving work groups.

When paperwork processing flow improvements are implemented, the next step is investigating the benefits and costs of automated data exchange with suppliers and customers. Electronic data interchange is a complex topic with rapidly changing standards and capabilities.

Most common standards for EDI are called ANSI X.12 (American National Standards Institute), and the still-under-development AIAG standard (Automotive Industry Action Group). Major vendors of third-party services (electronic mail boxes, and interface provision between different kinds of operating systems) include McDonnell Douglas and GEISCO (General Electric Information Services Company). EDI implementations can be lengthy and complex, depending on their scope and content, and the number of parties involved. However, their time saving abilities make them well worth the effort in many cases. There are a number of sources of information about EDI including user groups, professional organizations, newsletters, and EDI system vendors. Among them are:

EDI News
Phillips Publishing, Inc.
7811 Montrose Road
Potomac, Maryland 20854

EDI Executive
EDI Strategies
1225 Johnson Ferry Road
Suite 230
Marietta, Georgia 30068

EDI Report
Input Research
1280 Billa Street
Mountain View, California 94041

The EDI efforts in accounting will be connected closely with similar efforts in the purchasing area. The data transmitted will be different in most cases, but many sources and third-party functions will be the same. It makes sense to coordinate the overall EDI implementation effort so that it corresponds in time and leadership.

The area of management information systems (MIS) will find in the JIT environment different kinds of information are needed, the data will be analyzed and aggregated in different ways, the timeliness of data communication will become increasingly important, and the communications mediums often change. For example:

- Different kinds of information will be tracked and managed. For instance, product travel distances, percentages of activities which add value, housekeeping data, and cost reduction data.
- Different kinds of aggregation will be done. For instance, purchasing will want to measure quality, delivery, price, etc. at a commodity level. Manufacturing will want to measure percent of time which is setup, throughput times, etc. at the product line level. Management may want to measure productivity at a cell and/or focus factory level.
- Data communication timeliness will become more critical. As it becomes more flexible to meet end-item demand, manufacturing may find it needs to recalculate work load balances, manning levels for cells, and the number of kanbans required more frequently.
- Communication mediums sometimes change, as more information is transmitted via EDI, communicated by bar code, or triggered by (sometimes electronic) kanban.

Also, because of the need to utilize information quickly and effectively, information required to manage JIT operations can be understood and used far more effectively when presented graphically. The old adage that "a picture is worth a thousand words" is certainly true. In the opportunity assessment area, in the context of monitoring performance to goals, and managing day-to-day operations, the timely and effective communication of appropriate graphics is a lifesaver. Consider the examples in *Figure 14-1*, *Figure 14-2*, and *Figure 14-3*. Other areas typically launching improvement efforts coinciding with the initial pilot lines are purchasing, design engineering, and distribution. In all these areas, information obtained during the opportunity assessment should provide the criteria by which to decide what areas will be addressed first.

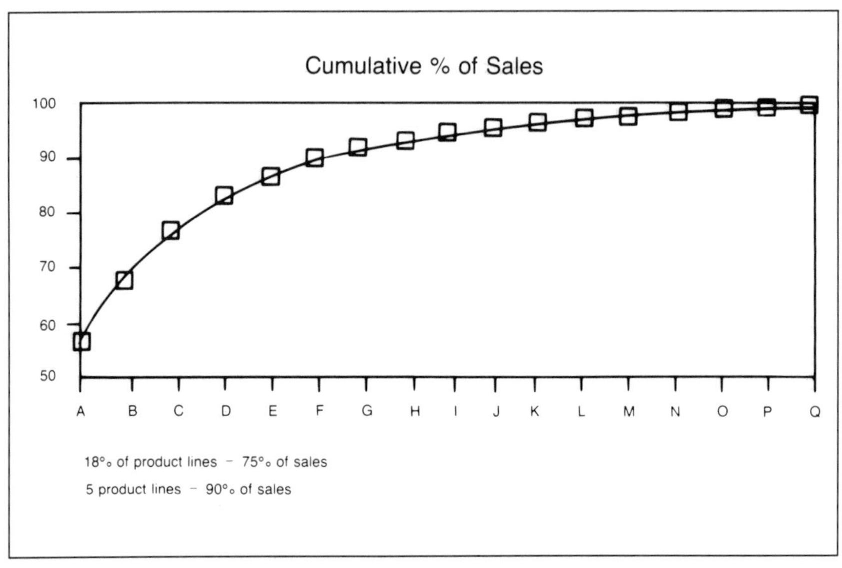

Figure 14-1. This figure illustrates that it is possible to improve those parts which account for 90% of sales by addressing the flow of only five product lines.

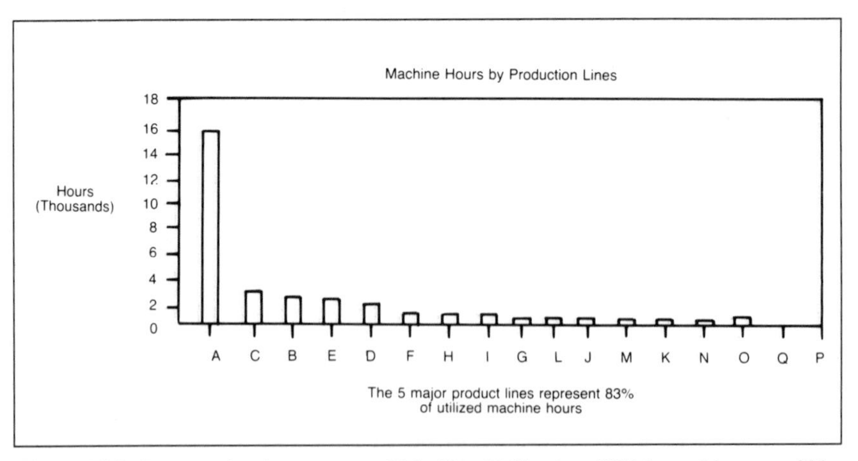

Figure 14-2. Opportunity Assessment of Machine Utilization. Which machines would be your initial candidates for preventive maintenance?

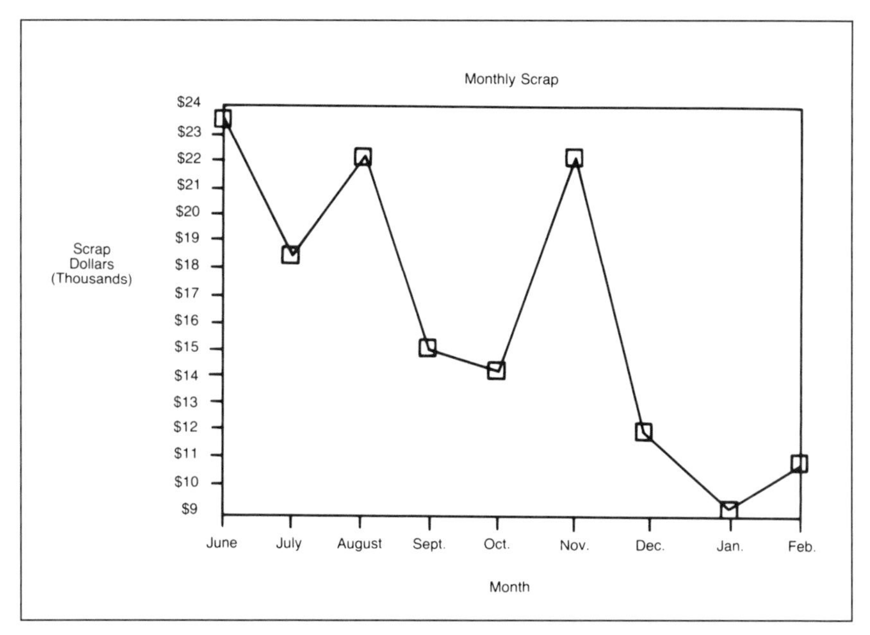

Figure 14-3. Ongoing Management of Quality. This figure illustrates impact of SPC and consistent focus on quality problems over one year.

In purchasing, efforts will begin by monitoring and managing P.O. volumes, and overall cost/delivery/quality performance by commodity. Suppliers should be brought into the facility and educated about the JIT program underway, and its impacts on the existing supplier base. One or two pilot commodities should then be chosen, based on criticality to pilot manufacturing operations, apparent supplier support levels, and the appropriateness of the buyers involved. An interdisciplinary team, comprised of purchasing, quality, design engineering, and manufacturing should then be assigned to work on this commodity. Objectives of the team's efforts should include:

- Narrowing the supplier base for the entire commodity to one or two suppliers.
- Consistent improvements in quality and delivery. (Attainment of certified status for all suppliers of the commodity).
- Implementation of frequent, point-of-use-delivery.
- Full utilization of comprehensive contracting techniques, including long-term capacity oriented contracts.
- Ongoing joint cost reduction programs with savings shared between supplier and manufacturer.
- Utilization of EDI for data sharing.
- Utilization of a pull system for material release authorization against blanket purchase orders.

Improvements launched in design engineering should include an initial thrust to streamline overall paperwork flow. The activities surrounding engineering change order processing are often fertile areas for JIT waste elimination, and should provide opportunities for relatively short-term gains.*Figure 14-4*, (*Figure 14-5*, and *Figure 14-6*).

Teams should be established to address manufacturability and quality in the pilot project design. After this effect is under way, candidates will often become obvious for cost reduction programs like VA/VE and CPR.

The distribution function may be handled in a couple of ways. Distribution of only the pilot project product may be addressed first, in which case beginning the effort at this stage of the program is appropriate. Another approach is to wait until at least the major product lines (in terms of sales dollars and production volume) have been addressed by JIT manufacturing implementation efforts. Because of the cross-product line scope of distribution activities, it requires fewer iterations of streamlining efforts when the bulk of the product lines are considered simultaneously. In distribution, JIT waste reduction efforts are usually managed best by distribution channel and/or geographic region. In either case, a team of marketing, traffic, customer service, quality and manufacturing people should be assembled to attack throughput times, quality problems, and excessive costs occuring in the distribution process.

Preparing for Remaining Facility Implementation

As pilot projects and the various support systems become operational, it will be important to prepare a "roll-out" strategy for remaining facility conversion activities. One of the most critical aspects of this preparation is identifying and training people to lead the new initiatives.

It is often easy to identify good candidates in group problem solving sessions, and in subsequent resolution implementation. They will be people whose opinions are respected by the rest of the group, and who are counted on to consistently oversee implementation tasks. The reliability and responsibility of these individuals, together with additional training, group dynamics and technical areas, will prove invaluable in later implementations.

When these individuals have been identified, replacements may need to be found to backfill their positions in the ongoing pilot. The replacements will need to be trained and assigned. The people removed to participate in remaining conversion activities should then be allowed to participate as appropriate in the selection of people for their implementation team, and in detailed implementation planning.

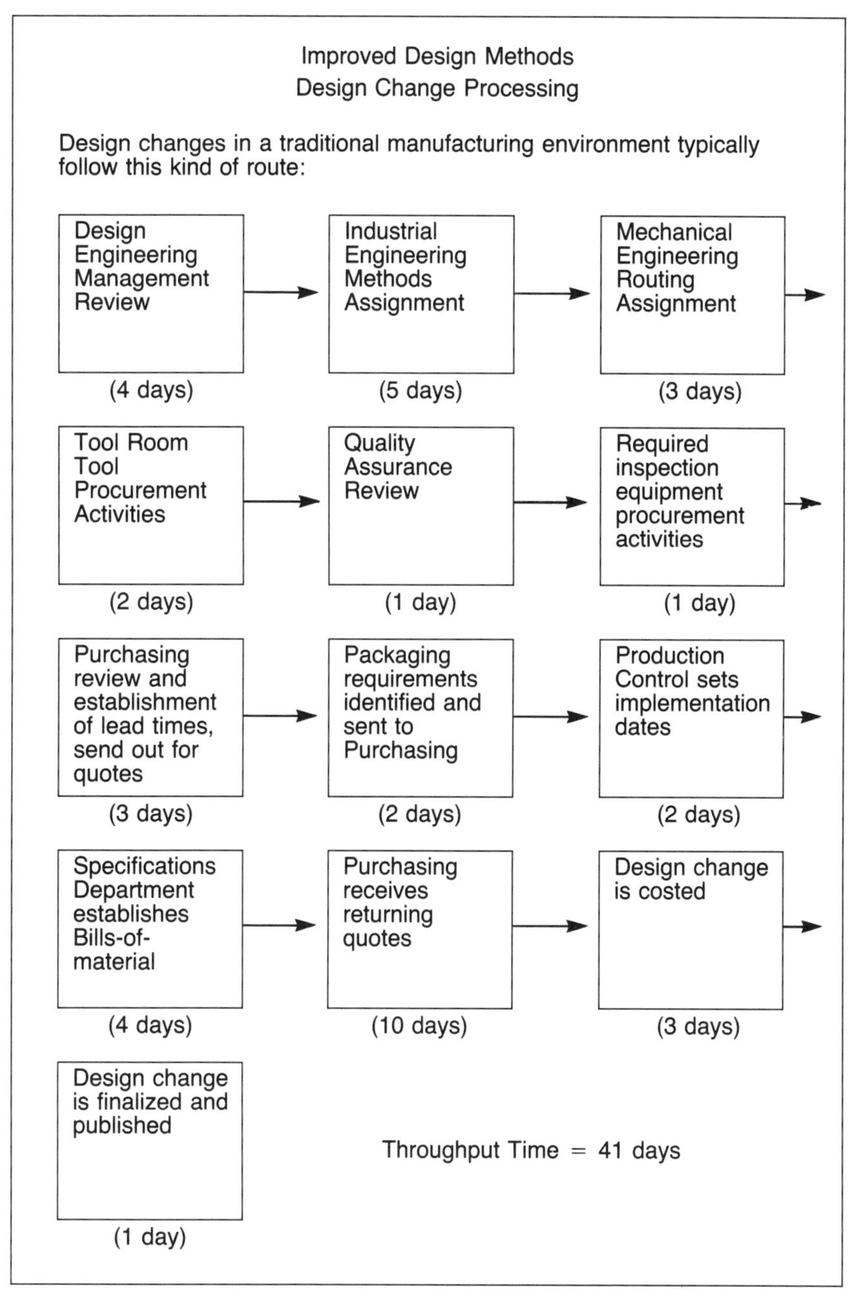

Figure 14-4. A typical pattern of design change implemenation within a traditional manufacturing environment.

Improved Design Methods

With the application of the Just-In-Time techniques, this process can often be redesigned in the following ways;

- Standardization of engineering change notice (ECN) formats, so that ECNs are more quickly understood and processed
- Elimination of most approvals, copying, and redundant handling
- Relocation of various departmental ECN processors into a single area to promote communication, reduce ECN travel distance/time, and improve visibility of work load imbalances
- Combine some job responsibilities such as Production Control and Purchasing reviews
- Design simplification to increase standardization and reduce the quantity of design changes required
- Adoption of "blanket" or "cover" ECNs to reduce analysis required for the establishing of effectivity dates

Figure 14-5. Improved Design Methods. Applying Just-In-Time techniques can redesign the process in a manner indicated in this figure.

Improved Design Methods
Design Change Processing

Applying these techniques, the following kind of flow can frequently be constructed:

Release from Design Engineering

Throughput Time = 16 days

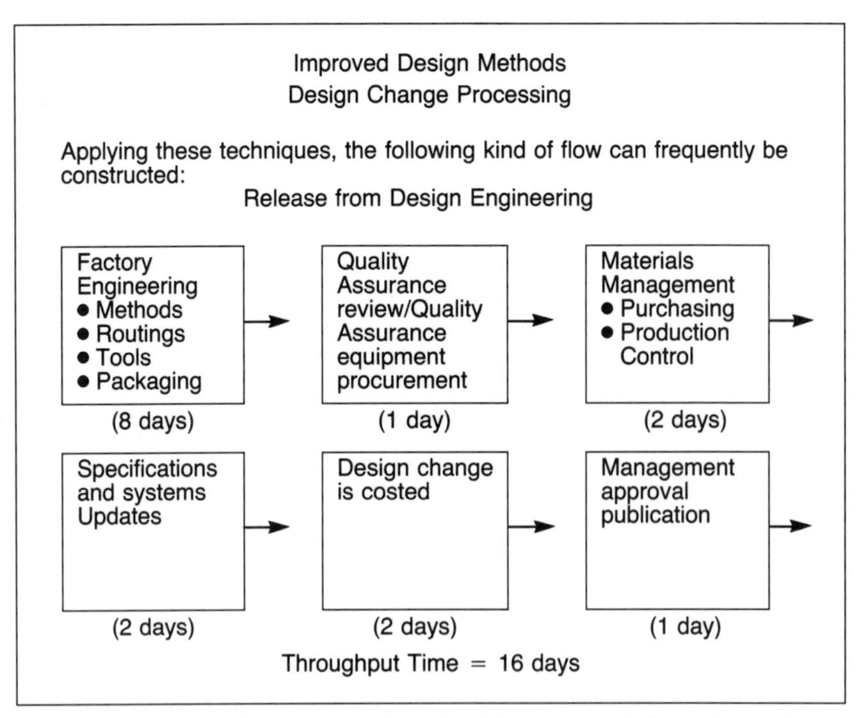

Figure 14-6. Design Change Processing. Applying these techniques frequently can achieve a throughput time reduction.

REMAINING FACILITY PLANNING AND DESIGN

15

The Remaining Facility Planning and Design segment of the JIT implementation program involves six steps; reviewing pilot project results; constructing a detailed layout for the balance of the facility; updating the original general implementation plan; creation of a detailed implementation plan for conversion of the balance of the facility, and development of specific performance measurements for the balance of the facility. This step completes the design and implementation plan for the future, becoming the detailed blueprint of the facility as it will be.

Pilot Project Review

When constructing the design and implementation plan for the balance of the facility, it will be essential to document and incorporate the lessons learned during pilot project implementation. Data to be considered include learning curves, levels of resistance by management in terms of spending and resource allocation, organized labor response, actual vs. expected improvement levels in all areas, time requirements for project personnel, demonstrated floor space requirements, and a number of others. Problems which arose and may recur as additional areas are converted, should also be noted. The point is not to repeat mistakes, and to incorporate any important experience from pilot project implementation in the design and planning of the remaining facility conversion. Management will generally forgive problems which arise and are resolved once. When the same problems return and become more prolific in remaining facility conversion, however, the support of senior management can dissolve quickly. Program leaders can't afford to lose the momentum which has been built up during pilot implementation at this critical juncture.

Detailed Layout for Remaining Facility

If the pilot project was properly selected, a wealth of detail should be available after it is underway about exact dimensions of equipment, required material handling devices, effectiveness of pull signal types, actual worker efficiency in a cellular setting, etc. All of this information must be incorporated into the design of the balance of the facility.

The blocks of floor space in the Macro level design generated earlier will need to be converted to individual work center layouts depicting all equipment, materials, and product flow.

Driving the design down to this level will probably require the assistance of mechanical engineering, industrial engineering, and plant layout personnel. Technical issues such as ventilation, lighting, sound, etc. will need to be considered. Also, because of the size of proposed changes, a more thorough cost/benefit study may be required at this point to justify planned expenditures for equipment relocation and equipment purchases. Based on feedback from the technical experts, experience with the pilot project, and management's level of support and/or funding, adjustments should be made to the Macro level implementation plan. It will be important to maintain the integrity of this plan so management can monitor overall program progress, and accurately anticipate returns on the invested resources.

Detailed Implementation Plan for Balance of Facility

When the detailed design has been completed, the steps used to create the detailed implementation plan for the pilot project can be repeated to generate the plan for conversion of the remainder of the facility. Differences will exist in the support areas, where the systems changes should be largely operational, and it becomes a process of transferring the new accounting methods, MIS functions, and production planning/control technologies to more product lines and/or areas. In addition, there will be more to take into account due to the volume of physical equipment movement, equipment purchase, and facility modification volume. It is best to generate a plan for the entire balance of the facility at this point, but construct it in phases, corresponding to the ''checkers-type'' movement referred to earlier. This way, individual steps may be handled as distinct projects, and remaining facility conversion becomes more manageable. Also, management can be encouraged to approve and monitor the overall program while not incurring a one-time surge in expenditure with questionable control and long-term ROI. A program structured this way is easier to sell, and easier to do.

Remaining facility conversion plans vary from six months to 36 months. The time frame depends on the number of projects involved, the implementation resources available, the priority of the conversion activities, and general commitment levels on the part of everyone involved.

Performance Measures

The final aspect of remaining facility planning and design which must be performed is the development of specific performance measures. One thing many organizations learn during pilot project operations is adjustments need to be made in performance measures that were originally established. Not everything measured was appropriate, some critical aspects of improvement were not monitored at all, and other measures have no reliable and consistent source of measurement data.

When adjustments are made, the original position should be benchmarked, so there is baseline data against which to measure future progress. Keep in mind the individual performance measures developed, when considered as a whole, should support the overall business objectives generated from the strategic business assessment.

With a detailed design, a detailed implementation plan, and appropriate performance measures in place, a solid foundation exists for remaining facility implementation.

REMAINING FACILITY IMPLEMENTATION 16

Implementing JIT operations in remaining areas of the facility involves assigning personnel, providing any required specialized training, initiation of the implementation plan, completion of support system conversions, and identification and resolution of problems arising during implementation.

Assignment of Personnel

By the time pilot operations are underway, there is often a backlog of people who have expressed an interest in participating. Successful pilots generate enthusiasm for the program, and many people want to get involved in order to share in the success. In addition, JIT operations, because of their orientation toward employee involvement and continuous improvement, provide an opportunity for employees to exercise some control, and create positive change in their own environments. For many, this is a more powerful incentive than wage increases. Leaders for the remaining conversion projects should already be selected and trained at this point, and encouraged to participate in selecting their own team members. Obvious limitations will be involved here, including job classifications, skill levels, etc. Management may have other considerations which also affect these decisions.

As the number of implementation projects or phases grows, smaller percentages of ''desirable'' employees–those highly skilled in appropriate areas, and possessing enthusiasm and understanding will diminish. The training and education required will become more challenging, and improvements may become less dramatic. In addition, the last areas converted tend to be those with the least overall potential for improvement, since projects are prioritized to maximize return on invested resources. At the same time, however, there is an increasing percentage of the worker population which is involved in and enthusiastic about the ongoing JIT program. Therefore, peer pressure and informal education will be more prevalent. In some settings, it makes sense to move people between cells and/or focus factories periodically, after initial implementations have been completed. This reduces the natural tendency to perceive groups as good or bad, superstars or remedial. The transfers should be done selectively, so general ''team'' spirit, which can be valuable, can be allowed to develop.

Provision of Specialized Training

As team members are assigned, special training needs often arise. Depending on cell structure and degree of cross training desired, special schools may need to be developed in-house to guarantee an adequate supply of qualified workers. When in-house training is not reasonable, agreements can be made with local colleges and/or technical institutes to provide the training required. It may be possible to obtain local, state, or federal funding to support these efforts.

A desirable criteria to use from the standpoint of flexibility and manning variability is extensive employee cross training. It is not uncommon to find wage scales in JIT operations which monetarily reward workers for the number of different work stations they are qualified to run. More flexible workers implies more and better training. It is that simple.

Initiation of Remaining Facility Implementation Plans

When team members are selected and trained, it is time to initiate remaining implementation activities. Since leaders in each area should be veterans from previous pilot projects, a sound basis of experience will exist in each group to guide ongoing activities. However, each new area will have its own at least slightly different equipment, people, and methods. As a result, it will be important not to stifle new ideas and innovations by attempting to avoid "reinventing the wheel."

Implementation activities should be reported by team leaders to the steering committee on a semimonthly or monthly basis. The steering committee should monitor the progress of the teams, and report it in terms of the Macro level implementation plan to senior management.

Completion of Support System Conversions

Initial support system development and installation should have been performed in conjunction with pilot project implementation. As more projects or phases of JIT implementation become operational, they must be converted to the new methods of accounting, design, purchasing, scheduling, etc. In some cases, the conversion of support systems will not advance as quickly as related manufacturing projects might. For example, it will probably require longer to incorporate quality and manufacturability into product designs than to implement improvements on the shop floor. Likewise, narrowing supplier bases and moving to comprehensive contracting is usually not a rapid process. Therefore, a "phasing in" period for these systems beyond the physical implementation time frame is necessary.

As with any other system conversion, there will be a cutover period involved, so that system validity and timeliness can be verified before the old procedures are discontinued.

Problem Identification and Resolution

During implementation, a number of operations and support systems problems will arise. Basically, the problems will be one of two types:

- Implementation related problems
- Ongoing operations problems

Implementation related problems involve how operations and/or support systems are being developed. They usually have a potential impact on the effectiveness of the future operation on a broad scale. Accordingly, they should be formally identified and resolved by the implementation team. Examples include problems with the way labor data is gathered, problems with the closing of all manufacturing (production) orders, and discrepancies involving job classifications and pay scales.

Ongoing operations problems are problems which, while they may have become visible as a result of the JIT implementation or even been triggered by it, are related to the operation itself. They may be relegated to quality circles, the employee involvement program, or normal resolution by management. Examples here include specific product quality problems, overloaded work centers and chronically late supplier deliveries.

Regardless of the problem, it is important to ensure that all of the problems are assigned to someone for resolution, and resolution is monitored.

PROGRAM MONITORING AND CONTINUOUS IMPROVEMENT 17

The major activities involved in the program monitoring and continuous improvement phase are overall program administration, transition of program format, and periodic review/reconceptualization.

Overall Program Administration

The general administration of the Just-In-Time program is usually handled by a program director, supported by the Steering Committee. See *Figure 17-1*, *Figure 17-2* and *Figure 17-3*.

Individual implementation projects are directed by project leaders, who typically become area, cell, or focus factory leaders as initial implementation activities are completed. Depending on the structure of the organization, the orientation of management, and the political environment, cellular concepts may be applied in support areas of offices as well as the shop floor. Because of the interdisciplinary nature common to office area cells, the roll-out of project leaders into cell leadership can be a more challenging process.

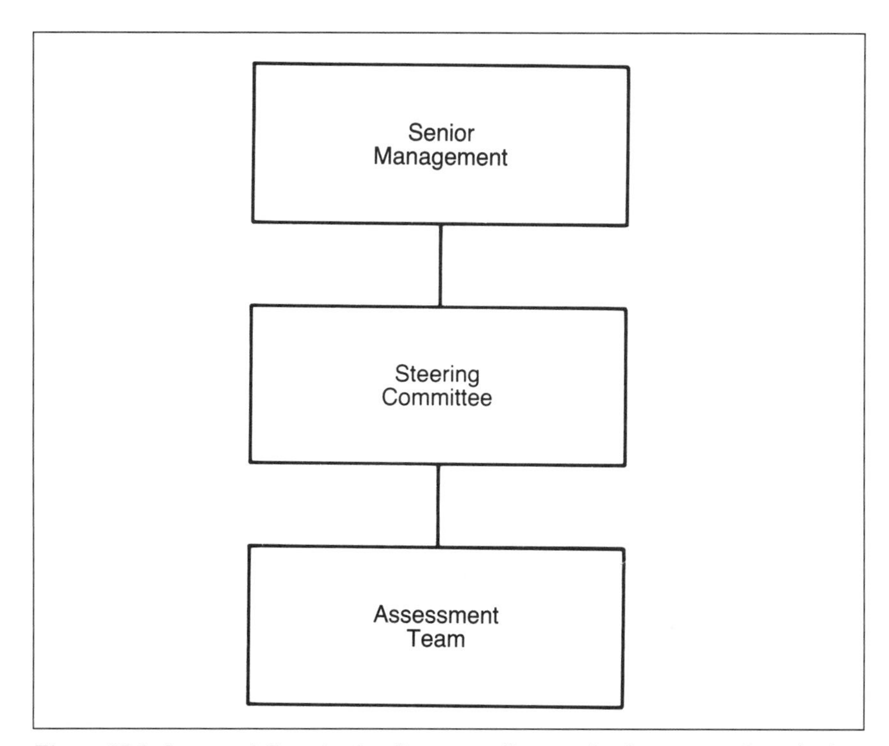

Figure 17-1. Suggested Organization Structures: Opportunity Assessment Organization.

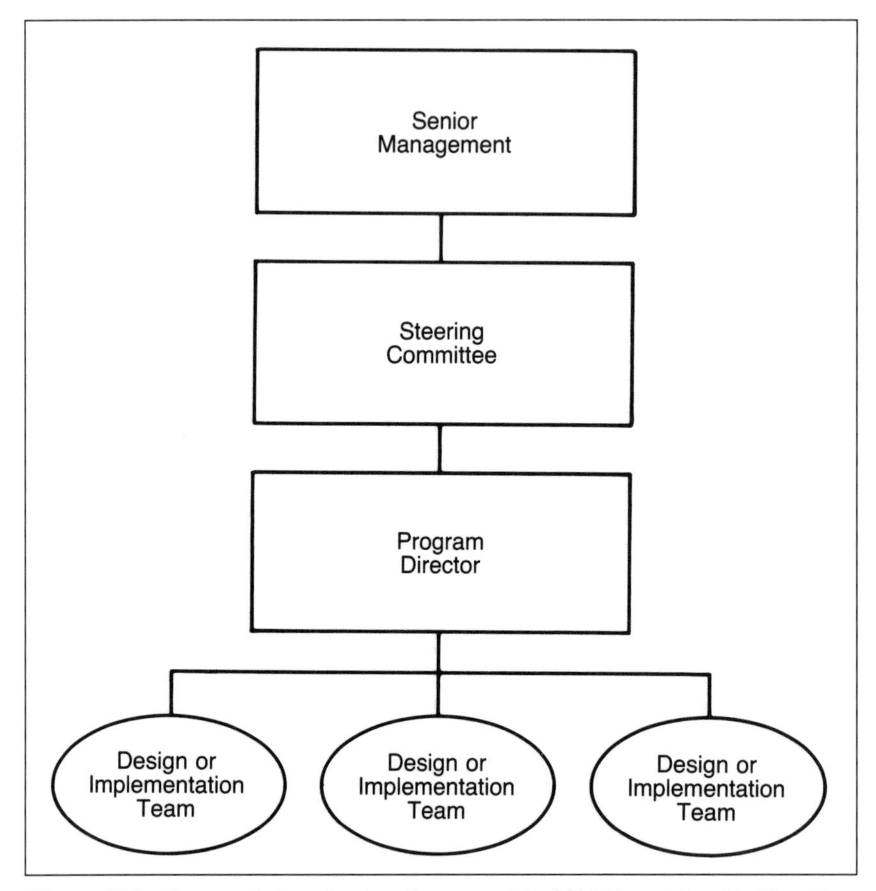

Figure 17-2. Suggested Organization Structure: Pilot and Remaining Facility Planning/Design/Implementation Organization.

The program director monitors ongoing implementation activities, communicating needs and problems to the steering committee as they arise, and accumulating normal reporting information for semimonthly or monthly meetings. The program director recommends actions to the steering committee regarding roll-out activities, personnel assignments, etc. The program director also coordinates ongoing implementation projects in the manufacturing and support areas, so pertinent data is shared between teams.

Transition of Program Format

As implementation projects are completed, the operating and reporting format of the program will need to change. The program should evolve into a continuous improvement mode, away from the rapid and major change approach used to drive the initial implementation.

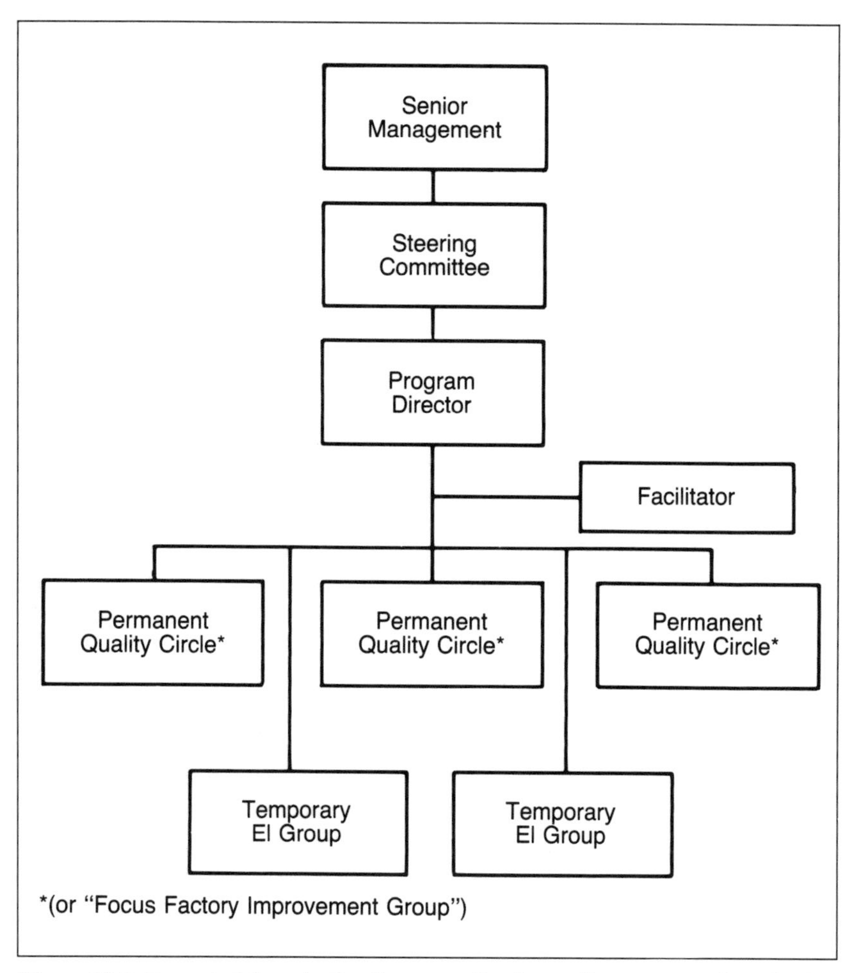

Figure 17-3. Suggested Organization Structure: Continuous Improvement Organization.

The major forces involved in the continuous improvement process will be centered in employee involvement groups and quality circles. While many of the principles of these two groups are similar, they operate in two distinct fashions, and usually attack different kinds of problems. See *Figure 17-4*.

Permanent groups, indigenous to a specific geographic area, continuously identifying and addressing operational problems in that area, are called "quality circles", or in the case of focus factory structures, "focus factory improvement groups" (FFIG). These groups collect problem data on an ongoing basis, and meet as time is available or at a minimum of one hour per week to prioritize and attach the problems encountered. Often, the problem data being collected and prioritized problem lists are displayed on blackboards somewhere in the area involved. As they are resolved, the problem is removed from the board, and the next one is attacked. These groups may be comprised of direct labor or management employees, or both.

The differences between the quality circle/focus factory improvement group and EI groups is illustrated in *Figure 17-4*.

Periodic Review and Reconceptualization

Just as every company should go through a strategic planning exercise at specified intervals, the company should also reconceptualize its vision of future operations. In many companies, this is an annual process which generates a five-year corporate plan. Because significant changes will occur in product mix, market size and makeup, manufacturing technology, distribution channels, and environmental factors, it is wise to re-evaluate specific manufacturing operations periodically as well. This is a process of reiterating the opportunity assessment activities involved in Chapter 12.

As American manufacturing organizations continue to evolve, more flexibility will be required to meet increasingly diverse and volatile demand. The company must continually review and realign itself to respond to these changes. After the initial JIT implementation, change is usually resisted far less, and the organization finds it less threatening. The required changes are usually much more quickly and easily incorporated as well, since the structure of operations has become more modular, and learning curves are reduced.

	EI Groups	Quality Circles
Length of Existence	Temporary, until problem is solved	Permanent
Focus	A specific problem	Any problems which occur in daily operations
Area of Operation	Could be anywhere	Limited to their own geographic area
Meeting Times	Regularly scheduled	As time permits, hopefully no less than one hour per week
Participants	Selected by steering committee as needed	Workers in a specific area

Figure 17-4. Characteristics of E.I. vs. Quality Circles.

INVOLVING THE EMPLOYEE 18

The employee involvement (EI) process is one of the most potentially valuable things a company can do. On the other hand, it can also be very distractive if not properly administered.

Many EI programs evolve from employee suggestion programs, and often meet with limited success. Even the most well-intended suggestion programs can alienate more people than they reward. This usually results from two weaknesses:

1. Suggestions are not well-understood by program administrators, and so, frequently not given adequate support.
2. Suggestions are often evaluated and rejected by area supervisors or managers who feel threatened by them, and the people who generate them.

Not alienating employees is essential to a sound EI program. Only when employees are certain that the program is fairly administered and genuine in its desire for new ideas can it succeed.

Many day-to-day operational suggestions may be incorporated via the quality circle. More significant problems or suggestions can be dealt with through EI. The EI program should attack problems which cross departmental or disciplinary boundaries. EI groups will, therefore, be comprised of interdisciplinary groups. Because they are oriented toward specific problems, these groups are typically disbanded at the completion of a project.

Employee involvement is a program fed by a continuous stream of suggestions or problems from employees. The suggestions/problems are accumulated and periodically prioritized.

Task groups are organized and assigned. The groups define the problem, brainstorm about underlying causes, collect data to verify what the major causes are, and develop potential solutions. Solutions are prioritized and evaluated, and a recommendation is formulated.

The recommendations are reviewed by the steering committee and approved. When approval has been given, joint responsibility for implementing the solution is assigned to the project team and supervision from the local area involved. Progress is reported to the steering committee until implementation is complete. At that point, the steering committee disbands the task group, and moves on to the next problem.

As the general JIT implementation projects are completed, the scope of suggestions or problems addressed by EI typically widens to include the activities which would have been handled under JIT implementation in previous months. Throughout times, setup reductions, work content imbalances, quality problems on specific purchased commodities, and particularly difficult-to-manufacture items are viable EI targets.

However, these targets can be adequately addressed only when management thinking has been realigned. Management which formerly believed they must control each problem themselves, will need to recognize it is impossible for them to deal with all problems. Where people were discouraged from voicing their problems, concerns and ideas, they must be encouraged and solicited for problems, ideas, and opportunities. In environments where business data was closely held and "secreted" away, workers must be viewed as partners in the business and kept abreast of serious developments. In short, management/worker relationships which were adversarial will need to be converted to environments where people function as effective idea-generating, problem-solving teams with common goals and objectives.

Effective employee involvement programs will build commitment on the part of all employees, improve response times in problem resolution, increase overall effectiveness of positive changes, and enhance productivity and morale. The EI program provides a model for positive change comprised of leadership, knowledge (or methodology), and employee will (motivation, energy). The underlying strategy for employee involvement consists of:

- Employee education (management and workers);
- Communication or expectations;
- Utilization of trained and proven leaders;
- Recognition and rewards for improvement, and
- Utilization of successes to foster further EI activity.

The EI organization will find that teams typically go through fundamental, evolutionary stages while becoming strong enough to deal with tough problems. These stages include:

The *building* phase, in which groups will be very tenative and uncertain about each other and their roles. Little cohesiveness will be exhibited, and it will be very difficult for them to develop a consensus.

The *developing* phase, when specific task related work is taken up by the group. All group members are involved in the resolution process and focus is narrowly targeted on discrete activities.

The *optimization* phase, during which group members prioritize and perform tasks on their own, come to decisions in a way which reflects consideration and respect for all the group members, and focus on underlying causes/relationships/processes rather than symptoms and characteristics.

Expansion of the EI program will be a function of adequate levels of leadership, understanding, enthusiasm, time/resources, and facilitating manpower. The rate of development, then, will be proportionate to the organization's ability to develop leaders and facilitators.

With these principles in mind, it should be clear that control, direction and support will be required on a continuous basis.

We have discussed the mechanics of the EI program in Chapter 10. The other significant aspect of employee involvement is the culture change required to support participative management efforts.

Dick Lounge, the Production Manager at GenRad, Inc., presented his view of the employee aspect of JIT in the SME Technical Paper titled *JIT Manufacturing: The People Aspect*. This material was originally presented at EMTAS '86 (March 17-19, 1986) in Los Angeles. Mr. Lounge's perspectives on the roles of various disciplines provide valuable insight into the culture change aspect of EI, and so his material is included here.

Much has been said about JIT in seminars and written in journals and articles. The first impression the unfamiliar reader envisions is that of trucks, filled with material, pulling up to the manufacturing facility, daily or even hourly with just the right amount of material for the day's production; but, in a greater sense, JIT is so much more than that. It touches all people and all functions throughout an organization in a most dramatic way. It requires a completely different mind set and a change in philosophy and strategy, that is totally opposite of traditional manufacturing teachings that have been taught in our universities and practiced in our plants for a very long time.

Indeed, the most difficult part of getting a JIT program started is convincing the people of this new philosophy, as a viable method of increasing productivity, reducing floor space and WIP to bring more to the bottom line. The people aspect is truly the most formidable problem to be solved and the most difficult one. A corporation's culture can be deeply engrained in its people and so too, can a traditional manufacturing concept be deeply engrained in people.

To change, adapt, convert or buy into a JIT philosophy, requires a grass roots movement by the people. To develop this grass roots movement, requires an education and communication effort, that is extensive and thorough with top management commitment and continuing support. Without this effort, it is very easy to fail, no matter how worthy the cause.

This paper describes concisely, the necessary training and educational topics which must be covered to successfully implement a JIT program, by obtaining a grass roots movement committed to success. The pitfalls are many and although management may give you full support, it will not work as a push down system. It must come from the bottom up with complete and total dedication that it can be done and it will work.

MRPII deals with planning and JIT deals with process. A consensus is beginning to emerge among management experts that JIT is a total commitment to excellence in manufacturing. That excellence is achieved primarily through the elimination of waste.

Inventory

Inventory is one of the first targets of the war on waste. Zero excess inventories is the goal of JIT. To help accomplish this, we need minimum lead times, minimum order quantities, and minimum safety stocks. There are two types of inventories, sitting and moving. Moving inventory can be reduced by process changes such as better material handling, quick die changeover, lower setup times and group technology to reduce product cycle times.

Sitting inventory must be reduced by improving the planning process, including marketing forecast, purchasing, and vendor deliveries, the stockroom, work in process and queues. Queues are a reactionary response to poor planning. Long or short queues are a form of sitting inventory that must be addressed as waste from the standpoint of JIT.

Production managers are rewarded for throughput, productivity, and other improvements, but not for reducing queues and lot sizes. Viewing queues as waste, we must overcome these attitudes of acceptable levels of queues.

Safety stocks are another opportunity for inventory reduction. Safety stocks are maintained because things are unpredictable. We tend to try to build coping systems instead of solutions. Reducing inventories in some areas while maintaining safety stocks defeats some of the benefits derived from JIT. Safety stocks are often nothing more than programmed waste.

The JIT effort goes beyond inventories to zero defects, zero setup, zero lot excess, zero handling and zero lead time.

Lead time is made up of setup time, run time, wait time, move time and queue time—70% to 80% of most manufacturing lead time is queue time.

Setup costs are best addressed by techniques such as, group technology, that attempts to simplify process and routing. Changing to a new product mix or options requires considerable effort, to focus on reducing this changeover time.

Lot size is also critical to setups. Smaller lot sizes mean more setups. It is desirable to produce one for one based on orders. This places a heavy emphasis on proper scheduling and it also implies some restrictions on ordering.

There is a trade-off between inventory carrying costs and the cost of changing the setup to accommodate another product. Purchasing decisions are often made to take advantage of volume price breaks. The extra costs may outweigh any advantages the discounts offer. The discounts are very visible, however, and the effects of carrying the inventory are not as readily apparent.

At certain times, volume discounts are significant. The key is to avoid being forced to accept volume deliveries, which would increase inventories and concentrate on separate negotiations for volume pricing and delivery schedules. Long-range planning will provide input on total volume that should allow the volume discount to be locked in.

Lot sizes should not be reduced unless the system is capable of supporting the change. This means the paperwork and stock picking system, as well as the physical system must be in place.

Setting production schedules on a daily basis will mean scheduling materials on a daily basis as well, even if this material initially comes from the stockroom instead of directly from a supplier.

The role of materials people is obviously key to any JIT program. It can also be very frustrating for these people, because the very measures which have been used to assess their performance will change.

Purchase price variance will become less critical as more emphasis is placed on delivery schedules and the time the items are actually in inventory and not being utilized in product build. The purchasing agent is also very concerned about establishing relationships with a single supplier where the norm before was to add a little protection factor by multiple sourcing as many parts as possible.

Changes to schedule are also a major concern to buyers, because this seems to be a fact of doing business and a reality in today's world. No one really believes that there will be a firm schedule that won't change. These changes are very traumatic for a buyer and take away their former feeling of security with multiple sources, volume discounts, and schedule changes being the norm.

Clearly, a lot of planning and commitment have to be made with marketing, suppliers, buyers and material managers to assure a clear understanding of each role. Time fences must be established for no changes, 20% change, 50% change, or whatever numbers are appropriate. A good place to start is with marketing to get a good idea of product demand. This is the driving force that establishes all other schedules so it is critical to be as accurate as possible, yet, this is also the area that is sometimes difficult to forecast due to market demand, product mix and other general uncertainties. The ideal situation would be to schedule and plan only on sales orders, but this would be a rare luxery. In reality, at least some part of the planning must rely on anticipated sales and product mixes.

Once the marketing forecast has been established, the next step should be to host a "Vendor Day" at your facility. During this activity the key suppliers can have the opportunity to see your product and its various assemblies and components. A tour of the Production Facility and processes will help to clarify your specific product requirements and clarify any general questions about your product.

Now it is time to get down to details. Establish the length of the time line for your forecasted requirements based on the marketing inputs:

- How much of the time line will be firm?
- How flexible and at what percentage?
- How often will periodic updates be made?
- How will suppliers be measured?
- By on time delivery?
- Quality?
- Response to changes within the agreed upon time fences?
- Price?

After these details have been worked out people will begin to have a better feeling about JIT. The older things they felt more comfortable with are no longer there; but, they are a part of defining the new parameters and guidelines as a team with marketing and the suppliers. The task does not seem quite so awesome and risky as the first assessment. Of course the true test is yet to come as the first deliveries start to come in and the proof of the plan begins to work. There will be problems, but then, there will always be problems no matter what system is used and good people will work the problems.

The main point is, inventory will be down. Cycle time will be reduced and therefore the cost of producing the product will go down, right down to your bottom line.

Education

Education is key in the implementation of any change, whether it is TQC, JIT, or new processes. For transitions from one practice to another to be smooth, education must come first. Next, is employee involvement. People must be involved in the decision making process and the problem solution process.

People deal much more effectively with change, when they are actively participating in that change rather than having change forced on them. Dealing with something new and different requires a lot of upfront selling through education, participation and management commitment. We must drive out fear among employees to raise questions and foster teamwork and cooperation to achieve common goals that are clearly understood by all. We must reward teamwork and flexibility, while focusing on business objectives, not specialization of tasks.

JIT requires people to assume a new mindset, a different way of thinking. The way you run the business changes. The elimination of waste becomes a key issue and requires an all out commitment to eliminate all defects.

A communications network needs to be established that identifies problems as they occur.

Awards aimed exclusively at increased productivity or work standards must be abandoned in favor of cycle time reduction, reduced queues, and WIP reduction, all of which will contribute to reduced inventories.

The Role of Production Control

Production Control continues to play a key role in the scheduling of production. The goal is to concentrate more on a cycle rate for product and assemblies with more emphasis on accurate scheduling information to fit the cycle rate. Product mix and flexibility are integrated into the overall plan to assure the proper build to satisfy marketing demands. Deviations from schedule will be a fact of life, no matter how good a system has been developed. Production control must recognize the deviations and adjust production activities to get back on schedule as well as expedite and reschedule. This probably is an indication that the original schedule was a problem or the operation was not properly able to execute a valid schedule. Expediting almost always leaves a long-term disaster of the schedule to accomodate a short-term need.

The people in Production Control would concentrate on the following activities:

- Plan the "Final Assembly Schedule" in advance. It should be level and planned to build some of everything everyday.
- Develop a "Master Production Schedule" from the "Final Assembly Schedule." The "Master Production Schedule" is actually a summary in daily buckets of the "Final Assembly Schedule."
- Explode the "Master Production Schedule" to produce level schedules for fabrication and subassembly operations.
- On a regular basis adjust the actual "Final Assembly Schedule" to reflect the change in customer orders or production problems.
- Monitor the entire pull system to make sure it is continuing to perform the proper functions of:
  a. Providing the proper material in time.
  b. Minimizing inventory at all points in the Production System.

The people in Production Control must work together as a team in close relationships with Manufacturing, and all of the processes to implement the primary mission of keeping production on schedule.

Supervisory Considerations

The role of supervisors is still key to obtaining maximum performance. This role changes somewhat with JIT/TQC as the team members of a production unit may actually rotate the team leader function among all members of the team. This gives each person a more thorough understanding and exposure to the overall operation of the unit; however, some team members may not be willing to assume the role of a team leader and prefer a more passive role as a key worker with no supervisory responsibilities. Difficulties may also occur when it is necessary to remind people to stay at their workstations, not to wander and chat. Punctuality issues may also appear, which deals with starting times, breaks, etc. The success of using a rotating team leader depends primarily on the maturity of the team as well as peer pressure to handle the routine supervisory activities. The paperwork requirements of a team leader will also be a problem for some people who detest paperwork and simply will not handle it well.

Training is also required to teach team leaders how to assume leadership roles, problem solving techniques and effective communication. As in any activity dealing with people and change, the groundwork must be layed carefully to fully explain the requirements and expectations. Working with the people as a team to fully establish the norms and goals will go a long way to assuring the success of the team.

In some team units where there is considerable resistance to assume the role of team leader by the team members, it may be more advantageous to establish a permanent team leader, who the other team leaders respect and can perform this function. This is usually better than forcing the team leader role on some people who will fail at this function because of their total resistance to it. Again, this is part of the overall philosophy of JIT to be flexible and adapt to whatever concepts work best for you, that will still accomplish your primary goals. The key concept here is that the people should have the opportunity to participate at whatever levels they feel comfortable with.

Once these issues are resolved we can concentrate on the day to day supervisory activities that will make JIT successful.

- Keep inventory low enough so that everyone has to concentrate on the process and how to improve it.
- Keep equity of work. No one should have the luxury of relaxing while others struggle to improve the system.
- Fine tune proposed inventory levels to reflect current problems or improvement status within the department.
- Work with Industrial Engineering to keep data and evaluate the quantity and type of people necessary to maintain different cycle rated.
- Maintain rebalancing for standard schedule changes or process improvements.

- If the card system is used, withdraw cards as necessary to keep the system tight.
- As much as possible, be involved in planning, training, development, process improvement, etc.
- Recognize the importance of the workforce in JIT and work to develop and improve workforce involvement.
- Forget about the traditional ''warm and fuzzy'' security blanket of management and supervisors (Plenty of Just-In-Case inventory to keep the workers busy and improve efficiency).

Major Factors Which Limit Growth

Key People Resource. A core group of people trained in a variety of tasks is essential to provide leadership and training to new people, who will be hired to meet our increased demand. The workforce should also be cross-trained to be able to perform a variety of tasks so that a flexible, adaptable group can respond to a multitude of task assignments. Most of our people tend to be too specialized in a particular task or product line to be effective, when demand in another area requires more resource quickly. If we are in a high growth mode, it is a pretty good assumption that others in our industry will also be experiencing some growth and there will be a great demand for people resource. This is why flexibility and cross-training must be stressed with our current workforce.

Material Availability. When we are in a high growth mode along with much of our industry, the demand for materials to build our products exceeds the available supply. The ability of manufacturers to fill the supply requirements is also limited to existing process capacity with a long lead time to bring on new process capacity. In this mode of operation our distributors and their suppliers go on an allocation system for the available parts.

In our traditional purchasing agreements, we concentrate on the lowest cost for a specified quantity of parts. In the future, we must also negotiate for allocation priority even it it means a cost increase.

Finals Area JIT Experiment

Our JIT experience at GenRad began in our final assembly and text work center on our PSP (Portable Service Processor) Test system. This area was very visable with a high probability of success for JIT strategies. Teams were formed which were named Q-PIT teams (Quality, Productivity Improvement Teams) to work on the process problems of production. Each problem was seen as an obstacle to success and an opportunity to improve our operations. KANBAN areas were established to hold a fixed amount of material, in our case, three PSPs, to work on. Storage shelves that formerly held work in process were disassembled and removed. Now there was no place to hold material except the designated KANBAN squares.

The work area was rearranged into a bench set up with a ''U'' shape. The finals test area was set up the same way. Then we allowed the work to flow. An amazing thing happened.

People began to communicate.

As soon as a problem was found, the people made it visable. They discussed the problem among themselves and set out to solve it. The problem had to be solved, because there was no extra inventory in WIP to draw from. The team took time at the end of each day to discuss the day's activities and made all the people in the work center aware of the problems that were uncovered and their solutions or progress. As each problem was solved, the work flowed smoother and smoother. A true sense of job satisfaction and accomplishment prevailed at the work center and people felt that they had control.

The following charts (*Figures 18-1, 18-2* and *18-3*) show some of the results that were obtained in this experimental area.

Pre-JIT	17.67 Days
4-20-85	11 Days
5-6-85	8 Days
8-1-85	8 Days

Figure 18-1. Lowered Cycle Times.

PSP Systems Work In Process	
Pre-JIT	16 Average
4-20-85	12 Average
5-6-85	8 Average
8-1-85	6 Average

Figure 18-2. Q-pit "Demand Pull" Team. Finals Area Experiment.

1984 Q4	50% Yield
1985 Q1	92% Yield

Figure 18-3. Improved Quality yield with the effects of the Q-pit "First Pass yield" and "Demand Pull" Teams, yield has improved by 42%.

Conclusion

People involvement at all levels is the key to implement any change, especially, Just-In-Time manufacturing. The commitment must be made by the people, because it is the people who will be most affected by the changes. Their day to day lives will not be the same and they must be able to understand this and feel comfortable with it.

Once you have the people committed, there is no stopping the revolution. Changes will occur routinely, as new ideas to eliminate waste and improve productivity will continue to emerge. This is a beautiful and rewarding experience and it makes Just-In-Time manufacturing a fun thing to do.[1]

JUST-IN-TIME AND FLEXIBLE MANUFACTURING SYSTEMS 19

A key aspect of management progressiveness is the willingness of a company's top level management to investigate new techniques for increasing productivity. When the choice is between Just-In-Time and Flexible Manufacturing Systems, management is faced with a number of considerations. Author James. W. Branam addressed these questions in detail during Autofact '87 in a paper titled *JIT versus FMS—Which Will top Management Buy?* The paper is reproduced in this chapter.

The italicized material in the parentheses in this paper has been added by the author of this book. It offers additional information from Mr. Duncan's point of view.

Introduction

The stereotype of a conservative, risk adverse, top management has been cited as the major cause of the limited productivity gains that have been obtained by many companies (*as we saw in Chapter One of this text*). MRP has not provided the level of productivity gains that have been attributed to Just-In-Time (JIT) and Flexible Manufacturing Systems (FMS). If companies are to move forward into new techniques should they move into JIT or FMS? How will a conservative, risk adverse, top management react to these two approaches? Why would they prefer one over the other?

Are each of these techniques just new fad programs or are they going to be an ongoing part of the manufacturing environment? Are these programs solutions looking for a problem or do they solve a real need?

The answers to these questions along with the major management consideration factors involved will be presented. Some of the factors that will be covered will be:

1. Cost
2. Potential Benefits
3. Employee Reaction
4. Potential Risks
5. Management's Perceived Probability of the success or failure of each of these approaches.

What Is The Problem?

Statistics have shown that the majority of time that a part spends in the factory (batch manufacturing) is spent in move and queue. The amount of time that a part actually spends on a machine is not all value added. The majority of that time is spent in non value added activities such as set up, tool change, and in process inspection. The portion of a part's throughput time in a factory that is spent in true value added activity is very small indeed. This percentage has been shown to be less than 2%.

The reason that the true value added percentage is so low is that the majority of manufacturing engineering, corporate and university research efforts have been spent trying to reduce the time involved in this value added portion. The basic functional scheduling, setup, tool change and in-process inspection efforts were left to production schedulers, foremen, material handlers, expeditors and set up men. These people had little, if any, formal training and even less professional help. They were thrown into the vast sea of the manufacturing plant and, in essence, told to sink or learn how to swim. Human nature prevailed and most of their efforts were survival oriented rather than being spent trying to solve the basic problem.

The advent of computers, Manufacturing Resource Planning (MRP) and exotic scheduling systems did little to alleviate this situation. The most basic and most difficult management task in a manufacturing environment is to balance schedule (or priority) and capacity on a continuous ongoing time phased basis. This basic task must be performed at two levels.

First, schedule and capacity must be balanced at the planning level. The schedule portion is typically performed starting with the Business Plan proceeding on through the Production Plan to the Master Production Schedule and on to the Material Requirements Plan. The capacity side of the equation also starts at the Business Plan level and proceeds on to Resource Requirements Plan, Rough Cut Capacity Plan and Capacity Requirements Plan. It appears that many companies spend a great deal of effort planning the schedule side of this equation but relatively little on planning the capacity side.

The second level of the schedule vs. capacity equation is at the execution level. At this level the schedule execution side of the equation is mainly concerned with priority control on the shop floor. This takes many different forms but typically involves a final assembly schedule, stock picking schedule, order priorities and operation sequence (or priority). The capacity side of the equation at the execution level takes the form of capacity control via labor control, inventory control, factory order control, machine (or work center) control, tool control, and preventive maintenance. The execution level is similar to the planning level in that one side of the equation seems to receive greater emphasis than the other. The difference is that it is the opposite side! The emphasis at the execution level is on the capacity side rather than the schedule side.

The net result of the heavy emphasis on planning the schedule and execution on the capacity side of the equation is that schedule execution usually leaves something to be desired. This weakness in schedule execution has been one of the biggest disappointments of MRP and other schedule planning computer systems. (*To some extent, as mentioned in Chapter One*), the development of JIT and FMS has been driven by the execution weaknesses of previous systems. These previous systems were long on planning and short on execution.

Just-In-Time

The "Just-In-Time" concept is broader in scope than just production and inventory control. It addresses the problem of productivity. The perception of the problem is that productivity suffers due to ''waste''. This waste is caused by inventory (too much) and quality (too low). In reference to quality it is said that U.S. quality has not become worse, but it has not become better either.

The Japanese define waste as anything other than the minimum amount of equipment, material, parts, space and worker's time which are absolutely essential in producing the product. This sound like good common sense; but how often is it followed?

The ''Just-In-Time'' concept is not—an inventory program, scheduling technique, for suppliers only, new fad, materials management project, cultural phenomenon, replacement for MRP, or a panacea for poor management. The objective of the ''Just-In-Time'' or ''Zero Inventory'' concept is not truly zero inventory but to reduce the inventory to the point where a problem is exposed, solve that problem and continue lowering inventory until the next problem is exposed, etc.

Many people tend to confuse the ''Just-In-Time'' concept with the mechanics that the Japanese use to implement the concept. The Japanese use the Kan Ban system to execute the ''Just-In-Time'' concept. However, the Kan Ban System works best in a highly repetitive, continuous production type environment. Kan Ban is much less effective in a make-to-order, job shop type environment. A significant point to remember is that the concept tends to apply universally, but the execution mechanics will have to be different based on the type of business.

The FMS Concept

A FMS is described as a group of machines clustered in a work cell that has automated material handling and will produce multiple parts using multiple process steps via computer control.

A job shop has multiple parts going through multiple operations, work centers, and departments. A transfer line has one part, going through multiple operations on multiple machines in a totally automated line (automated material handling). The development of numerical control (NC) machines started the trend of the job shop towards the transfer line concept. This was further advanced through the use of group technology work cells. The next (and present) step was to automate the material handling between these machines and to make them all NC Machines and to control all of these machines plus the material handling from a single computer. The debate continues as to whether FMS is really Flexible Manufacturing or is it a multi-part (or flexible) transfer line.

FMS is not ''totally flexible'' at this point in time. It is generally restricted by ''cube size'' and part configuration—round, flat, square, rectangular, etc. Therefore a plant operating totally on the FMS concept would have multiple FMS's to make all of the parts for a complex assembly. However, the concept of focused manufacturing might allow for multiple small plants each with a single FMS. This single FMS would produce a single type of product (within the same cube and same general shape).

The development of automatic transfer lines in the high volume continuous production industries and the development of numerical control (NC) machines for the job shop industries started converting the control of the actual operation time from the man to the machine. The further development of the group technology work cell concept started the switch from the specialized (by type of work) departments or work centers to the specialized (by type of part) work cells. The next step was the automation of the material handling between the machines in the work cell and the computer control of all of these machines plus the material handling between them.

The development of the FMS, the concept of taking a part from raw material to finished part in one work center and the lot size of one has offered the opportunity for a whole new concept of manufacturing. This new concept of manufacturing not only offers the opportunity for more productivity in the direct labor areas but in other areas as well. The simplified flow of parts that are processed in a FMS (raw stores to FMS to Assembly) allows reduced material handling, less complex tracking and cost control as well as more simplified ordering and scheduling. Besides reducing the labor and confusion mentioned above, the simplification of the flow will also reduce the size of routings and the effort to produce them. The significant advantages of reduced throughput time and reduced inventory that are provided by the FMS are greatly enhanced by the potential for improvements in other areas.

Lot Size Of "One"

While all of this automation was being developed for the purpose of reducing the labor content of manufacturing (the most difficult variable to control) another concept was being developed. This concept was the "Just- In-Time" or "Zero Inventory" concept. This concept was aimed at reducing both inventory and throughput time. It has been proven that inventory required is a function of throughput time. Therefore, if throughput time could be reduced, then inventory could be reduced as well as providing better customer service.

It has also been proven that the majority (80-90%) of throughput time is queue time. The goal of FMS is to also reduce throughput time which also provides more responsiveness to changes in customer demand. Therefore, these concepts started coming together.

The other major stumbling block to Just-In-Time, FMS and reduced throughput time, is set up time. It, therefore, became the goal of both Just-In-Time and FMS to attempt to drive the economic set up quantity to "one". If the economic set up quantity could become "one" then component fabrication batch sizes could be set based on any desired assembly batch size (based on customer demand) without a cost penalty.

Types of Companies

There are two basic types of companies: Make to Stock and Make to Order. The Make to Order companies further breakdown into two types: Assemble to Order and Engineer to Order. The Engineer to Order company is the true job shop. The major task of manufacturing management is to balance schedule (load) and capacity. Both the Make to Stock and Assemble to Order companies rely on adjusting the schedule (load) to conform to available capacity. On the other hand, the Engineer to Order companies must adjust capacity to balance to the schedule (load). This is obviously a much more difficult task. It is necessary because the Engineer to Order company is usually held to a delivery date based on a contract. This contract may, in many cases, have a penalty clause for failure to meet the contractual due date. This major difference between Engineer to Order (job shop) companies and Make to Stock or Assemble to Order companies must be considered when evaluating these systems.

The Make to Stock companies tend to be high volume repetitive type manufacturers. These companies have, in the past, made good use of transfer lines or other "hard" automation. They have also been well disciplined or regimented. This has helped them become successful with MRP systems. Now this type of company is trending towards flexible transfer lines (for more product variation) and/or JIT.

The Engineer to Order companies tend to be low volume job shops. The traditional functional manufacturing flow (mill dept. to lathe dept., etc.) generally prevails in these companies. These companies have relied heavily on the innovation of their people and thus have not always been as well disciplined or regimented as the Make to Stock companies. To convert from a functional flow to a process flow is a major change for the Engineer to Order Job Shop. This change is, in many cases, being accomplished via the use of Group Technology (GT) work cells and/or FMS cells. (*See Chapter Six of this text.*)

Another difference in the type of company that can impact as to which approach to choose is whether the company or plant is old or new. Older companies or plants tend to be more regimented and making major changes to this regimentation is very difficult. Also these older companies or plants tend to have stronger unions that make the introduction of new technology or techniques more difficult. Newer companies, on the other hand, generally have less regimentation and less powerful unions. This, combined with the fact that the employees tend to be younger and more interested in becoming part of the future trends, makes it easier to install new technology or techniques. These considerations are important. Companies that ignore the human element in making these types of changes are going to have a lot of problems. This point may seem so obvious that it goes without saying. However, it is amazing how many companies get enamored with the technology or techniques and completely ignore the human element. As a matter of fact some companies install automation because they do not want to deal with the human element.

The Answer

The answer to the question in the title is "It Depends." It depends on several factors. The factors mentioned above such as type of product and type of company are important. Also important is the previous experience of the company, its workers and most importantly its management. If a company could not make an MRP system work, then it is unlikely that the discipline or regimentation is present to make the Kan Ban type of JIT work. Many managers feel that automation is the only lasting answer to productivity improvements. They also feel that the JIT concepts are valid but that the mechanics are weak for the long term. The following sections will compare JIT and FMS on the five factors listed in the Introduction. These are: Cost, Potential, Benefits, Employee Reaction, Potential Risks, and Managements Perceived Probability of the Success or Failure of Each of These Approaches.

Cost

JIT costs less than FMS to start up. JIT requires very little capital outlay. Most of the cost is education and training. The education and training is necessary for both management and the workers. The required attitude change is both necessary and difficult. Customers and suppliers cannot be overlooked. They must also be educated. All of this education and training must be continued and maintained on a long term basis. Also any cost involved in making the business conform to the JIT type environment must be considered.

FMS and/or Computer Integrated Manufacturing (CIM) will cost substantially more than JIT. It has been estimated that U.S. companies will spend over $40 billion on CIM over the next five years. Worldwide the expenditure will be double that of the U.S. The FMS cost will also have to include education and training costs just as JIT did. However, the type will be different. There will be less emphasis on attitude change and more on the technical side. The large initial capital outlay is difficult for many companies to accept. The depreciation that follows the capital outlay will impact the bottom line profitability of the company for years to come. This increase in the fixed costs will no doubt tend to raise the break even volume of the plant or company.

Potential Benefits

It is estimated that the potential cost reduction benefits of JIT are as much as 50% to 80% of FMS. This compares to a JIT cost of maybe only 10% of the FME cost. Lead time reduction using JIT is also as much as 50% to 80% of that that can be obtained using FMS. Quality improvement would also fall in the same range of 50% to 80% of that of an FMS. Flexibility is probably greater with JIT than it is with FMS. This is due to the limitations involved in the shape and cube size that are built into the FMS. Overhead will definitely be lower with JIT. Customer Attitude should improve with both JIT and FMS and it may be difficult to measure the difference—if they are both working properly.

The potential for cost reduction with FMS is enormous. However, the business must be there to support the large investment that is required. It is technically easier to lay off an employee that it is to sell a machine or an FMS (without a loss). The cost reductions using an FMS range from 30% to 60%. The FMS cost reduction is usually greater than the JIT cost reduction but it is not in proportion to the costs involved. Lead time reduction of 30% to 60% is also possible. The lead time reduction is also greater than JIT but again not in proportion to the costs involved. The potential for quality improvement is very high with FMS. Quality is presently rated as the key item that is required for the U.S. to regain its competitive position in the world market place. There are fewer inconsistencies in FMS quality than with JIT quality due to less human intervention. However, many experts have said that the attitude change that is required for JIT through education and training is also required for FMS or CIM to make quality the first priority. The zero set up time and a lot size of one makes the FMS very flexible—within a range of shape and cube size. However, the increased people flexibility of JIT while not offering the speed of flexibility that FMS offers, does offer a greater range of flexibility. There is no question that FMS has a higher overhead factor than does JIT. This is due to the higher capital investment and depreciation as well as the higher level of support in the area of hardware maintenance and software programming. If the business volume is great enough, then the greater overhead will be more than offset by the lower cost of production offered by the FMS. Customers may be more impressed by FMS on plant tours and the faster flexibility than by JIT. However, they may be more pleased with the greater range of flexibility and the quality attitudes of the JIT environment.

Employee Reaction

Many unions resist JIT because:

1. Potential Employee Reductions
2. Increased Flexibility Required by the Employee
3. The fear that Partipative Management will undermine the Union's Strength

Many employees question if management really will continue with participative management or just practice it until the next crisis comes along. Some employees see JIT as a way to save some jobs without running the risk of closing down or of extensive automation that would cost even more jobs. However, this could possibly just pospone the inevitable since the long term trend of JIT is also toward automation but on a more evolutionary basis. The Job Enrichment aspect of JIT is appealing to many employees. It should also be noted that some employees are happy with routine and do not want to have to think on the job.

Unions will resist FMS but will eventually lose or acquiesce on the subject of automation. Automation has been the American way for years. Employees will fear the loss of jobs and will resist to some degree. Some employees will attempt to adapt and try to become a part of the new FMS approach. FMS is the automation step that will finally drive manufacturing the way of the family farm. That is, from labor intensive to almost totally automated and thus requiring far fewer people than in the past.

Potential Risks

The first potential risk with JIT is that the attitude change education and training will not "take." Lip service in this area will not do. There must be a strong management commitment and involvement as well as a strong program of education and training. The American people tend to resist strong discipline and/or regimentation. JIT is somewhat contradictory in that it involves more participative management but at the same time it requires more discipline and/or regimentation. The program must be maintained on an ongoing maintenance of the discipline and regimentation. Perhaps the biggest potential risk is management itself. They must be committed to the participative management process and must continue with it if JIT is to be successful.

The first and most obvious risk of FMS is that the technology will not work. The newest portion of the FMS technology, the automated material handling (robots, AGVS, etc.) appears to be the biggest problem at the moment. The large cost of FMS is a potential risk from two standpoints: (1) Getting the Money to Start; (2) The Ongoing High Overhead Due to Depreciation.

It is important that the business volume be maintained to support the large inventment. If there is any doubt then there is risk of spending the company into bankruptcy. Technology is changing fast and will continue to do so at a faster rate. The risk of jumping in first and getting some early profit or of waiting for better technology and possibly losing market share in the meantime is a big one.

Management's Perceived Probability Of The Success Or Failure Of Each Of These Approaches

Regardless of all of the logical objective criteria mentioned above the true deciding factor may be management's perception of the probability of success or failure of each of these approaches. The management of fairly new (15-20 years old-maximum) high volume Make to Stock type plants should tend to buy JIT because the atmosphere is correct. Employee attitudes will usually be more favorable than with older plants and plants that are Make to Order. Unions will not be as firmly entrenched in the new plants. The high volume repetitive nature of this type of plant tends to lend itself to JIT. The management of older plants (Make to Stock or Make to Order) and that of most Make to Order plants should tend to resist JIT (at least in its purest form).

The JIT concepts apply universally but the mechanics do not. FMS uses the JIT concepts but applies them using different mechanics. The evolutionary direction of JIT is towards the use of FMS. Therefore both are going in the same direction. Going directly to FMS is tantamount to skipping a step. It may be necessary to skip the JIT step in some situations depending on the competitive situation.

When evaluating JIT vs. FMS it is necessary to make sure they are not just solutions looking for a problem. It is first necessary to determine the problem and then evaluate the techniques in view of how they would solve the problem. The problem in much of U.S. industry today is productivity (lack of it) (*as we discussed in Chapter One*). The problem of productivity is waste. JIT focuses on waste as its primary target. FMS is really an advanced application of JIT.

FMS brings automation and flow process technology to the job shop. Therefore the question is not which one to use but whether to take the intermediate step of JIT before moving on to FMS. The risk will be higher if the move is direct to FMS but this risk will vary by type of company. The necessity to make the move direct to FMS may be driven by competitive pressure.[1] (*The danger in this approach is that poor methods and sequences will be automated.*)

References

1. James W. Branum, "JIT versus FMS—Which Will Top Management Buy?", Autofact '87, 1987 Conference Papers, (Dearborn, MI: Society of Manufacturing Engineers).

JIT APPLICATIONS 20

An Overview

In this chapter, the reader is introduced to a few recent cases of JIT applications, focusing especially on the increasing combination of JIT with technology change (i.e. flexible manufacturing and robotics).

Applications are drawn from the writings of A. Thomas Jacoby (Eastman Kodak™), James W. Branam (Rockwell International), James W. Kelley (IBM), Richard Hammond (Ernst and Whinney), and from *Manufacturing Engineering* magazine.

Glenn Graham (Coopers & Lybrand) outlined some of the applications of Just-In-Time in the SME book *Automation Encyclopedia: A to Z in Advanced Manufacturing*. The following five paragraphs are from that book.

Just-In-Time in a general sense applies to any kind of production. Improving process capabilities, reducing setup times, and similar activities are useful to any manufacturer. JIT promotes repetitive manufacturing. Over time, manufacturers have transformed a job shop operation to repetitive production as their volume increased. Henry Ford made the transition with his first assembly line. This conversion means reducing the inventory between operations, transferring material in a flow, and decreasing the throughput time. One way to think of it is as an attempt to make an entire industry perform as if it were part of a single assembly line.

There is, however, a great deal to be learned from JIT which can be applied to a job shop. JIT has not been applied correctly to any industry, so it is a question of how smoothly a job shop can be expected to perform. Many manufacturers who operate by job shop methods may not need to do so. It is worth a review to determine if job shop production is potentially repetitive. Even if that review turns out negative, most of the principles of JIT apply except for uniform plant load.

There are valid reasons that require job shop methods. Some of them are: unique customer orders which require custom engineering; operations performed are "on the edge of the state of the art," which suggests that yields are poor and sometimes unpredictable; quality control requires inspection or holding in lot sizes; and small-volume orders arrive at irregular times. While a valid condition exists that demands a job shop operation, it may not always be true. A review of JIT principles may show a way.

JIT applies to a substantial part of the operation. JIT may apply to a combination of types of manufacturing. It is most fully developed when used to perfect production that is already repetitive. As with almost anything that is difficult, it is easy to disclaim applicability.

JIT requires a positive attitude, a determination to seek out problems and solve them, and nonacceptance of the belief that barriers to a smooth flow of production are permanent. Companies which have embarked on JIT have rarely attained all they wish to do, but they are much better off than if they had decided not to try it at all.[1]

Susan Lloyd McGarry (Yankee Group) outlined some JIT application advantages and discussed JIT and computer-integrated manufacturing (CIM) in her article in the May 1986 *Manufacturing Engineering*. The following nine paragraphs are particularly of note.

Challenging Tradition

The reason JIT has been able to challenge some of the assumptions of U.S. manufacturing is that JIT results in savings. For example, JIT has reduced inventory, and thus reduced carrying costs, by more than 50% for some companies. Other areas of savings include reduced manufacturing time as a result of shorter setup time; better work flow, resulting in a better quality product; and reduced space requirements (some manufacturers have reported space savings of more than 30%.)

JIT is also bringing about more cooperative relationships between customers and suppliers. The companies at the forefront of JIT are working closely with their suppliers to ensure quality and to help the suppliers implement JIT themselves. The JIT advocates are educating their suppliers, demonstrating the kind of cost savings that have been achieved internally and offering technical help to some of their most crucial suppliers. The goal is to make it a win-win situation so that both manufacturers and suppliers can take advantage of JIT.

As an example of this cooperation, many companies are sharing their build schedules with their suppliers, allowing the suppliers to plan ahead. One manufacturer reports that it freezes its schedule for 30 days, forecasts the next month with a variance of $\pm 10\%$, and forecasts the third month with a variance of $\pm 20\%$. The manufacturer's suppliers can then plan their own schedules and contact their suppliers to ensure shipment deadlines.

What does a company need in order to implement JIT manufacturing? First, successful execution of JIT requires longterm management commitment. Because the scope of JIT is large and threatens a number of accepted tenets, JIT needs to be company-wide and understood not as a program that will be abandoned in one or two years, but as a change in philosophy. Secondly, JIT is predicated on the idea of worker involvement. Workers in a JIT facility are viewed as experts in the processes they perform. This often requires a shift in management attitude, particularly among middle managers.

The implementation process can be divided into two components: ensuring quality of the product and ensuring quantity of the product. JIT assumes the part is good because a company cannot afford to have one bad part. Many companies adopt statistical quality control or total quality control at the same time or before implementing JIT. Most companies begin with a pilot project.

JIT and CIM

JIT manufacturing provides the necessary discipline for successful investment in computer-integrated manufacturing (CIM). Too often, automation equipment has been bought merely for the sake of purchasing equipment. Before buying solutions, companies must first analyze their manufacturing processes, determine how these processes can be made more efficient and decide on the goals of their automation strategies. By using the techniques included in JIT to examine shop floor and manufacturing processes, a manufacturer is much more able to make decisions regarding investment in technology.

At the same time, CIM compliments the JIT manufacturing approach. For instance, the initial savings from JIT come from changes in procedures; automating those procedures can result in additional savings. Further, CIM offers the kind of flexibility that will be required as manufacturing moves closer to lot sizes of one and setup times of zero.

Stated simply, investing millions of dollars in CIM without examining the manufacturing process will yield an expensive disaster. In today's, and more so in tomorrow's, competitive manufacturing situation, the 80% improvement achieved by examining the process will not be enough. To survive past 1995, companies will have to take the concepts suggested by both JIT and CIM and make them work in their environments.[2]

JIT with regard to flexible manufacturing was discussed briefly by Edward A. Herring (Digital Equipment Corp.) in his SME Technical Paper *Defining Flexible Manufacturing*. In this paper, he wrote:

JIT is neither an application nor a system. It is a concept with sets of cross functional goals to reduce this ''built-in'' safety net. The JIT concepts require greater discipline in all areas of manufacturing (planning, materials, maintenance, manufacturing and sales) than previously required for flexible manufacturing. With this high level of discipline, flexible manufacturing will operate successfully under the umbrella of JIT.[3]

Just-In-Time inventory control together with flexible manufacturing systems meant more cost-effective production methods at Eastman Kodak. A. Thomas Jacoby discussed how the two systems were successfully combined in *Manufacturing Methods Revamped at Kodak*. The article appeared in the May, 1986 issue of *Manufacturing Engineering*. It is reprinted in the next few pages.

The italicized material in parenthesis has been added by this book's author. It offers additional information from Mr. Duncan's point of view.

While the Eastman Kodak™ Co. is probably best known for its film and cameras, the firm also produces a wide range of products, including copiers and duplicators. From the standpoint of manufacturing, Kodak has adopted the latest techniques in flexible manufacturing systems (FMSs) and parts inventory in an effort to upgrade its production environment and narrow tolerance windows.

FMSs are a recent innovation. Modern mass production techniques have resulted in specialized machines designed to do one job in a highly automated, low-cost manner. Dedicated machines are most cost-effective when used for high-volume jobs, however, when their cost can be amortized over many pieces, over a long period. At the Kodak plant in Elmgrove, NY, such machines produce small items in quantities of up to one million per month. At the other end of the spectrum are parts that are required in monthly quantities of hundreds or thousands, and these items are manufactured in a more flexible fashion. The fuser roller used to apply toner to paper in a Kodak Ektaprint™ copier-duplicator is one such part.

Figure 20-1. One of two robots stationed at the twin lathes in the fuser roller cell is shown placing a tube on a conveyor.

Consisting of a milled aluminum tube layered with rubber, the roller design has been gradually improved over the years. Because the roller is just one of the designs used in Kodak's line of copier-duplicators, it doesn't justify a dedicated machine to produce it. Instead, the part is produced using a robot lathe cell that can be programmed to manufacture the tube and then quickly reprogrammed to produce another, completely different part that requires turning, milling, and burnishing. In fact, with the two lathes built into the cell, the station can simultaneously produce a roller on one side and another part on the other.

In operation, the anodized aluminum tubing is extracted from a hopper by a central robot arm. This robot can access a pair of hoppers and correctly direct the raw stock to either of the two lathes. The stock is removed from each lathe's carousel by a dedicated robot arm that mounts the piece in the lathe, removes it after machining, and then directs it to other stations within the cell for milling, burnishing, and washing. The process is automated to such an extent that a single operator can now handle work that formerly required four people.

Today, an increasing number of workers are employed in support positions, rather than in actual production of parts. This trend makes jobs more varied, and increased mechanization of production reduces scrap losses. For example, before the robot cell was installed, each of the steps required to produce the fuser roller was carried out in a separate location. Repeated handling often resulted in damage to the rollers, and this damage often was sufficient to scrap these labor-intensive parts. In the new cell configuration, robots are programmed to perfectly reproduce the same set of motions time after time. Once a robot is set up to perform its functions correctly, it will continue to function within set specifications for long periods.

In the past, there was at least one rejected part in every lot of weldment assemblies for the roller parts. Since the lathe/robot cell has been in operation, more than 300 lots have been produced without a single reject.

A key to successful use of FMS is the ability to quickly restructure a given production line to make a new or different part. Among the advantages of this approach is much faster part turnaround time. FMS reduces on-hand inventory requirements because, if supplies run low, new parts can be produced in a matter of days. One part that formerly required a 40-day turnaround from receipt of order to completion of the first part is now produced in just one day, eliminating the need for stockpiling the components and tying up floor space.

The best example of flexible manufacturing of parts for the line of Kodak copier-duplicators is the robot variable emission welding cell. This station requires no setup time, and on a cost per part basis, it makes no difference if it

Figure 20-2. This robot welder, designed by Eastman Kodak Co. can handle as many as 16 different jobs on one turn of its conveyor.

is used to produce one part or 1000. The station can go from welding one type of part to a completely different type of part with no setup penalty whatsoever.

Kodak uses a great many weldments in its copier assemblies despite their greater cost and complexity to produce. While castings are cheaper, they're heavier and more difficult to work with. Weldments are rigid and lightweight, and they retain their at-rest position better than screwed-together subassemblies. By eliminating the variables in positioning and welding setups, weldments can be produced with excellent precision and repeatable accuracy. The robot welder makes weldments with this high degree of consistency, and moreover, it never gets tired. In addition, it is flexible enough to produce a variety of different parts. Ultimately, FMS reduces the cost of parts and improves the reliability of products for end-users.[4]

The following case study on file with SME is indicative of the success that is becoming increasingly common through the merger of JIT philosophy and advanced technology/robotics.

Material handling is an area that can benefit from employing Just-In-Time techniques. Author David K. Doerflein offers an overview of the process flow with the flexibility of the gantry robots followed from barstock to painted parts. The concept was presented in a paper titled *Just-In-Time with Multiple Gantry Robots* at Robots 11/17th ISIR.

(Again, the italicized material in this paper has been added by the author of this book. This material should offer additional information from this book's authors point of view).

''We have a one and a half million dollar inventory setting outside, covered with rust. The part the customer needs yesterday is not available and won't be available for 30 to 45 days.'' Murphy's Law, it happens more often than we like to think about.

Let's solve the problem by introducing the ingredients of the problem.

a. Why do we have such a large inventory?
b. Why isn't the part we need in inventory?
c. Why will it take 30 to 45 days to manufacture a part?

Let's answer each question, arrive at good sound solutions but don't cloud the solutions with a multiple of ''what if's.''

We need to manufacture only the parts required in a given time period. That answers all three questions. Simple answer, but not a simple solution.

Solution Decisions

We must establish a manufacturing process that will produce parts, based on customer orders. We must then produce this daily order requirement while maintaining or increasing the quality of any part in a family of 500 different part numbers. This system has to be conceived without losing sight of the ''Just-In-Time'' principle.

A team of professionals would decide on the equipment required and list the best equipment available to perform their tasks. Keep in mind, we are looking at a lightly manned system to produce a high volume of parts as required per day. Also, this family of parts is presently being manufactured on high-volume transfer line machinery.

System Equipment

The operatons to be performed are:

a. Bar stock cut-off
b. Accurately located drill holes
c. Length, squareness, flatness and hole location checking
d. Notch cutting
e. Deburring
f. Heat Treat
g. Temper
h. Paint
i. Palletizing.

Each of the items listed above was reviewed individually as to its adaptness for being a flexible process or flexible piece of equipment. Once the decisions were made on the equipment requirements, an enormous material handling task was tackled.

Material Handling

The material handling tasks were outlined into the following categories:

a. Machine load/unload
b. Conveyor load/unload
c. Palletizing of tiering racks
d. Load/unload of transfer devices.

The demands that are associated with a flexible manufacturing system indeed were present in the handling system. In fact, the demands increased, due to the fact that manufactured parts from a second line of production were being introduced into a common handling system. The handling method was becoming more obvious and apparent.

A flexible method that would allow for a multitude of variables but have the ability to do simple repetitious fixed automation type moves. "A common structure with multiple carriage gantry robots controlled individually through a computer cell controller" would be a possible solution.

The wheels were rolling. Here was an opportunity to assemble a "start to finish" manufacturing system utilizing a myriad of processing and material handling equipment communicating with each other.

Process Description

The overall application will perform in the following manner:

Four individual gantry robots perform specified duties within a defined work envelope with slight overlaps of work areas.

The first gantry robot picks up steel forging stock from a CNC cut-off saw, and presents the part to one of two fixtures at one of two machining centers. A machined part is then removed from a fixture and placed onto a pick-up station, where the second gantry robot will pick up the part and present to either a coordinate measuring machine or a floor mounted six axis robot equipped with a plasma arc cutting torch. If presented to the measuring machine, a check will be made on the length, squareness of the cut-off and accuracy of the machining. Not every part will be inspected, therefore, most parts will move directly to the plasma cutting operation. As in previous operations, dual fixtures are employed and alternated for maximum efficiency.

The second gantry robot also moves the parts from the plasma cutting operation to fixtures located in front of two floor mounted six axes robots equipped with tooling for sharp edge removal. At these stations, the plasma cutting, machining surfaces and cut-off edges will be deburred.

The third gantry robot will remove the parts from the deburr stations and place them on a set-down station. Additionally, the third gantry robot will be unloading different parts from a wire guided vehicle, and placing them for pick-up by the fourth gantry.

The fourth gantry robot will be kept busy by performing conveyor loading duties. The unique task for this robot is handling the current manufacturing line of parts, while having the capability of one of five hundred (500) different parts being introduced from a second manufacturing line.

The conveyor being loaded in this area will accelerate the parts to a programmed speed conveyor that will carry the parts through a heat treat furnace.

At the present time, only one furnace is used in the system. This one furnace accounts for the necessity of having to funnel the second manufacturing line into the fourth gantry area of line one.

To continue this process, the parts must be quenched and then tempered. During the heat treat and quench operations, the parts are positioned side by side in two rows. To pass the parts through the temper furnaces, we will use a single carriage gantry robot as the material handling method. This robot arranges the parts into eight rows for temper conveying.

Following the temper process, the parts are set onto tiering racks or placed on AGVs. This operation is again being handled by an overhead gantry robot.

Parts placed onto tiering racks are used for ongoing production at this site or other machinery manufacturing plants. Parts placed onto automatic guided vehicles are routed to an automatic pain line and will be used to fill the repair parts requirement.

In this "Just-In-Time" concept, we have not lost sight of the original objective:

1. Reduce inventory
2. Produce only the parts needed for the customers daily order requirements
3. Every piece of machinery has the range and flexibility to automatically adjust for part changeover.

Communications And System Control

In this installation, all of the robot controllers are equipped with a communications option that allows unsolicited commands between the cell and robot controller.

Each manufacturing line has its own computer cell controller linked to a supervisor (host) computer.

All of the material handling robots, processing robots, machining centers, cut-off machine and coordinate measuring machine are connected to the computer cell controller through terminal servers by the way of ethernet and through a mechanism coordinator (programmable logic controller) via data highway.

This method allows feedback from several devices that are critical to robot decisions without placing the processing burden entirely on the cell controller.

The control system is designed to operate automatically. However, it is always under supervisory control of a system operator. The system operator has ultimate control of the availability of the system devices and components. Operating personnel will be prompted by the system control to perform certain functions and report their actions back to the control.

Robot Programming Enhancements

To reduce program maintenance and to reduce programming time, additional enhancements will be utilized.

In order for a duplicate part number to be manufactured on any manufacturing line, decisions had to be made, based on past experience. Machines, fixtures and robots cannot be positioned or duplicated close enough for a common robot program per part to be used across several lines. Therefore, a series of frame alignment tables relative to pick-up and set-down locations were established that could be merged with a common part program, thus shifting the program while being used at that particular robot controller.

To further reduce robot programming time, an off-line programming is being used via a personal computer. Since 500 parts will be programmed across 18 robots, 9000 programs must be generated. By off-line programming, complex moves or slight editing of existing programs can be accomplished without interfering with production.

The new or edited programs can be sent directly to the host computer for down loading to the appropriate control for production or testing.

Application Verification

To prove that an application of this type can be justified and indeed accomplishes it's goals, three separate system flow analysis were performed. A static analysis was run to determine the quantity of equipment required.

To verify this data, two separate dynamic model simulations were run by non-related parties. The results were amazingly similar.

In a "Just-In-Time" manufacturing system environment, utilizing gantry type robots as the material handling method and machine loading method, a significant number of specialty pieces of equipment are not required.

Additionally, a "Just-In-Time" manufacturing system can be as simple or as elaborate as is necessary or justified.

In conclusion, when concepting a "Just-In-Time" manufacturing system that requires a large work envelope, maximum utilization of valuable floor space and the dexterity of multi-axis robots, gantry or multiple gantries should be considered in your application.[5]

The publication *CIM Technology*, (Volume 5, Number 1) carried a feature which briefly describes the JIT/MRP relationship. (This case study from Dover reflects a company in transition from MRP to JIT orientation, as discussed in Chapter One). The article titled *Marrying JIT and MRP* is repeated in the next several paragraphs.

In the mid '70s, Dover Corp.'s Ohio Pattern Works Div. (Cincinnati) implemented a Material Requirements Planning (MRP) system. By late 1980, the company was realizing inventory reductions of 26%, increased productivity of more than 9%, backlog reductions of 78%, customer service improvements of 42%, and reductions in inventory purchases of 10%. Even so, Dover found it increasingly difficult to compete in an environment where distributor and original equipment manufacturer inventories had become non-existent, and instantaneous delivery had become the expected norm.

Enter Just-In-Time (JIT). The company examined the strengths and weaknesses of both MRP and JIT in the context of Dover's manufacturing environment. It found MRP to be a superior planning system and JIT to be a superior execution system.

The "courtship" of the two systems began in 1984. Dover shifted from the traditional system of material movement to the simple "bucket" system of JIT. Because this change altered Dover's operating methods only slightly, inventory balances and accuracies were not affected. Stockroom inventories, and their reporting remained identical to those of pre-JIT. The MRP system was manually altered, adjusted–even ignored–to compensate for the integration of a JIT execution system.

Since instituting this change in its assembly operations, Dover generated customer service levels in excess of 99.8% within two days of order entry. These results have been accomplished with a virtually infinite number of product variations and annual volumes of more than 300,000 units. Dover is now on its way to expanding the MRP/JIT system into its entire manufacturing arena, as well as to its key suppliers.[6]

An additional case study from IBM provides a more technical analysis of critical elements in the planning of manufacturing lines, with especially valuable insights regarding the application of robotics engineering in JIT manufacturing environments.

James W. Kelley's case study details how incorporating JIT and Continuous Flow Manufacturing (CFM) would have improved the design of a robotic manufacturing line. The remarks appeared in a paper presented at Robots 12 and Vision '88, titled *The Impact of JIT and CFM Concepts on Robotic Manufacturing Line Design*. The next pages are devoted to that subject.

It is difficult to move from one manufacturing methodology to another. To change, requires the people doing business as usual (BAU) to accept the risk involved with moving to promising, but unfamiliar concepts.

In many cases, massive organizations and related software systems have evolved to support BAU. When originated, these organizations and software systems were genuinely needed to expedite the manufacturing methodology of that particular era. For those practicing BAU, it is not unreasonable to view change with some reservation, expecially if new methodology questions the usefulness of existing organizations and software support systems.

The 1983 Robotic Line, described in this paper, questioned the usefulness of organizations and software support systems that controlled manual manufacturing methodology. Predictably, the BAU groups were hesitant to endorse a robotically controlled manufacturing line.

JIT/CFM proponents do not oppose the use of robotics and automation, but their guidelines for the use of these techniques are different from the old return on investment (ROI) test used in the past. If you are open minded, but polarized toward automation and robotics, your first discussion with a staunch JIT/CFM proponent will probably be both informative and defensive. Some robotic supporters will resist JIT/CFM concepts.

The goal of this paper is to show how the utilization of JIT/CFM concepts would have changed the architecture of the robotic line previously described, and to identify some potential JIT/CFM traps. To accomplish this goal, the following major topics will be discussed:

- The 1983 Robotic Line Architecture
- JIT/CFM Concepts
- JIT/CFM Improved Robotic Line Architecture

The 1983 Robotic Line architecture is described in detail in this section. Most of the material was taken directly from a paper presented at internal IBM Design For Automation Symposiums in Dallas and Munich.

Detailed descriptions are given, so that the reader will have a sound basis for comparison of the 1983 Robotic Line architecture to the theoretical JIT/CFM Improved Robotic Line discussed in the last section of this paper.

The Plant Automation By Sector Concept

The Lab Mechanization Dept. attempted to provide an overall system design architecture that could be applied to the entire manufacturing plant. Past experience at IBM had shown that it was extremely expensive, in terms of capital equipment and personnel, to totally automate a plant. Based on this experience, the Mechanization Dept. developed the plant automation by sector concept as the highest level of system architecture.

This concept employes the following sets of rules:

1. A sector is defined as a logical group of tools or precesses that produce parts.
2. Plants are automated, sector by sector, as new areas for cost effective automation are identified.
3. New sectors are not forced to accept old sector technology.
4. Sectors are self contained, including sector buffers and database, and are transportable from plant to plant.
5. A processor called the Sector Controller controls the flow of jobs through all tools in a sector.

6. A processor called the Plant Controller controls the flow of jobs through all sectors in a plant.

Plant/Sector Controller Job Flow

A job is defined as a record that specifies what processes or operations a part must undergo. Associated with a job i.d., job priority, job routing and job description. Jobs controlled by the Plant Controller include in their routing the sectors that jobs must be processed by. *Figure 20-3* illustrates Plant Controller job flow.

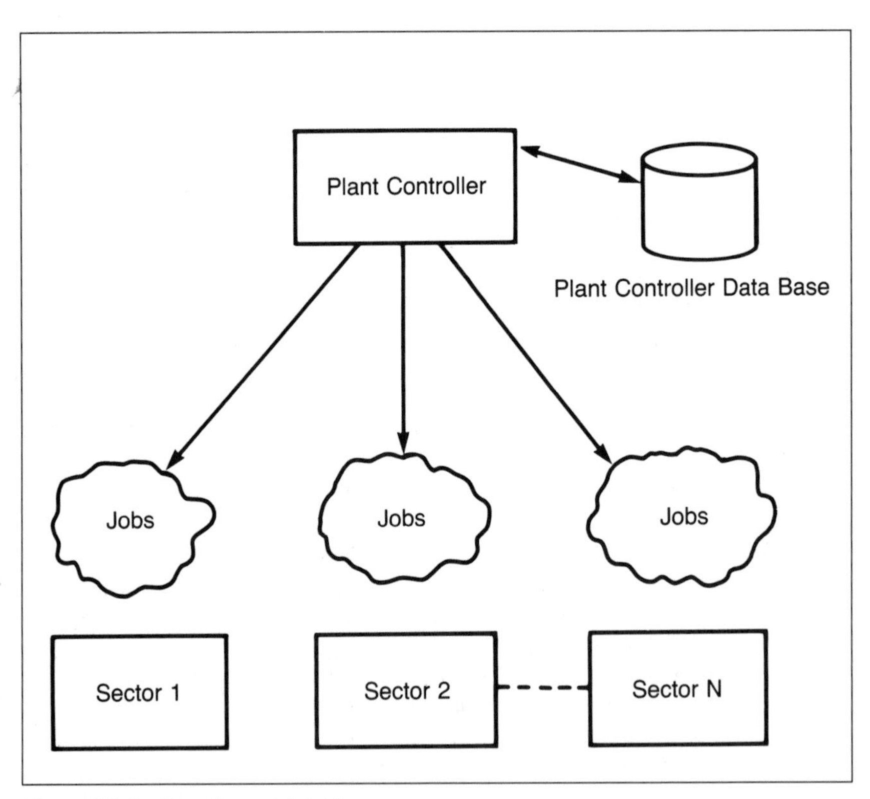

Figure 20-3. Plant Control Job Flow.

The Sector Controller works on a lower level where it deals with single tools or processes. Job routings for Sector Controllers include a list of operations associated with the tools. *Figure 20-4* illustrates the Sector Controller job flow.

The Benefits Of Plant Automation By Sector

Plant automation by sector provides the following major benefits:

1. Plant Controller failure does not bring down individual sectors since sectors have input/output and database buffering.
2. Plant Controller software remains stable because it is buffered from technology changes by individual sectors.
3. Entire sectors are transportable from plant to plant because they are self contained.
4. New plant technology is maintained because new sectors use the latest technology.
5. Individual tools/processes are optimized because the Sector Controller is designed specifically for their control.
6. Old plants can be automated on a sector by sector basis with minimum risk.
7. It is easier to understand the control required for a group of tools/processes than an entire plant.

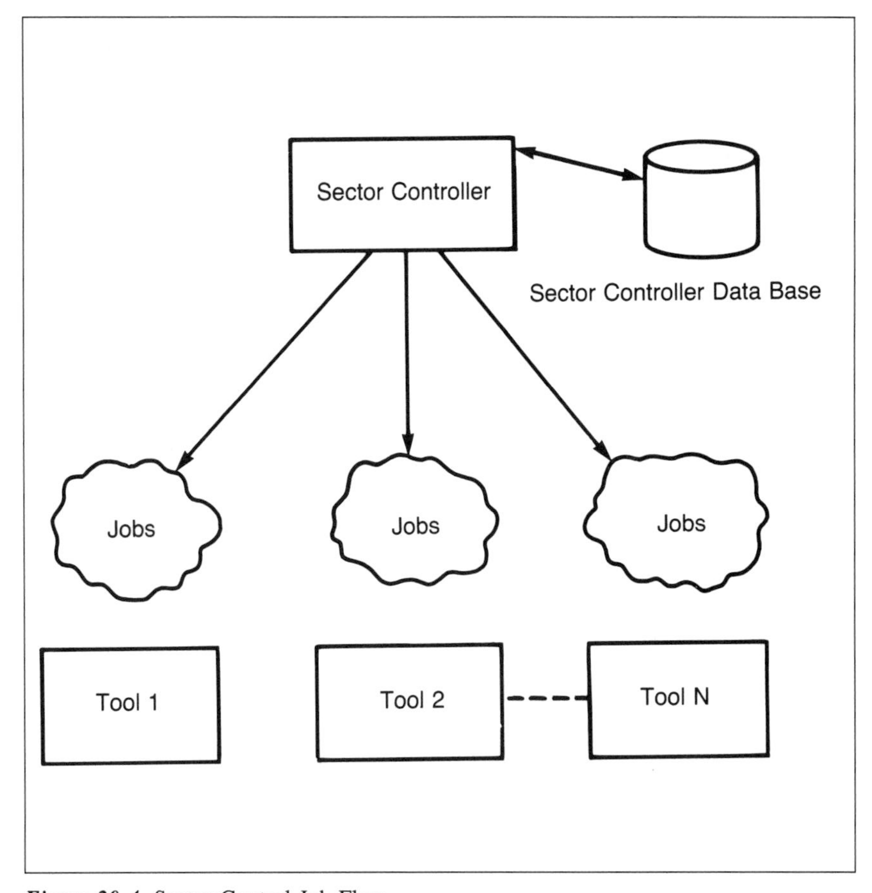

Figure 20-4. Sector Control Job Flow.

Part Movement/Buffers Defined

Before moving on to the description of the flow through the Robotic Line, it is necessary to understand what gets moved and how parts are buffered. A description of part movement/buffering follows.

Row/Carrier Movement

The smallest part handled by the Robotic Line is defined as a row. *Figure 20-5* depicts seven rows being placed on a metal carrier. The carrier is identified by a barcode i.d. and the row by an alpha numeric i.d. The job of putting the rows on the carrier and reporting the row and carrier i.d.'s to the Sector Controller is handled by a single mechanized tool that will be described, in detail, in the Tooling section. The metal carriers serve as part holders for the remaining Robotic Line operations and are individually processed by each tool.

From one to twelve carriers are loaded into magazines in preparation for tool processing. Each tool is equipped with input/output, and in some cases, reject elevators. Magazines of carriers can be placed directly in the tools input elevator or in an Input Buffer Tray. Although magazines have no barcode or alpha numeric i.d.'s, the Sector Controller creates a pseudo magazine i.d. that is used to track the movement of carriers through the line. Since tool operations take an

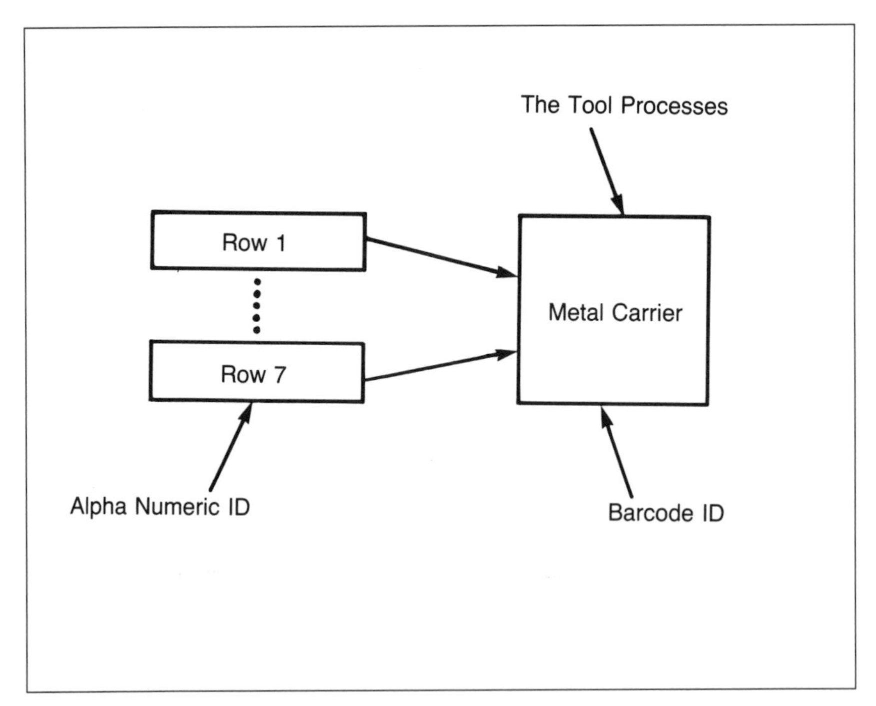

Figure 20-5. Row/Carrier Movement.

average of 1.5 minutes per carrier, a full cassette served as an 18 minute buffer for the tool. *Figure 20-6* shows 12 carriers being input to a magazine.

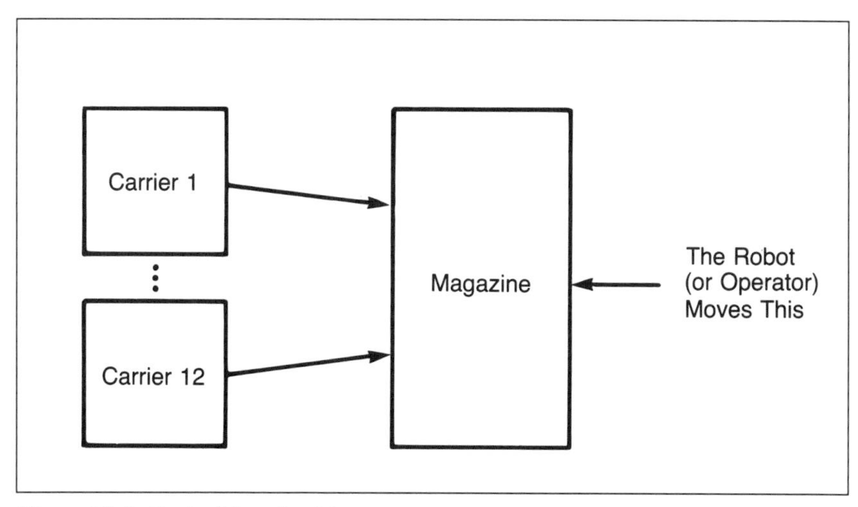

Figure 20-6. Carrier/Magazine Movement.

Magazine/Buffer Tray Movement

It would be inefficient to move single magazines of parts to and from the Robotic Line, so magazines are placed in Buffer Trays. The Buffer Tray depicted in *Figure 20-7* has a capabity of 16 magazines. Each tray has a barcode and operator readable i.d. Also shown in *Figure 20-7* is an example of job location. Job 1 is made up of four magazines and Job 2 of one magazine. Operators move the Buffer Trays to and from Buffer Tray Bins similar to the one shown in *Figure 20-8*.

Magazines of plastic carriers holding 7 parts each (MPC), magazines of empty metal carriers (MMC), and paper travelers (T) enter the line at the Tray Assembly Station (TAS). At the TAS, an operator palletizes the magazines into job groups and sends the job related data to the Sector Controller (SC) via terminal or barcode reader. The paper traveler does not move beyond this station.

After receiving job data and updating its database, the Sector Controller directs the movement of the palletized Buffer Trays (BT) to the Buffer (B) of Tool 1 (T1) or to the Input Buffer Tray Bin (IBTB).

The Sector Controller directs the movement of all parts through the line permitting a ''paperless'' mode of operation. Paperless movement control is accomplished in the following manner.

When tools require a magazine of parts to work, or a magazine of parts to be removed, they send a service request to the Sector Controller via the Communication link (C). The Sector Controller responds by directing either the Robot (R) or an operator to move the appropriate magazine and then verifies that this

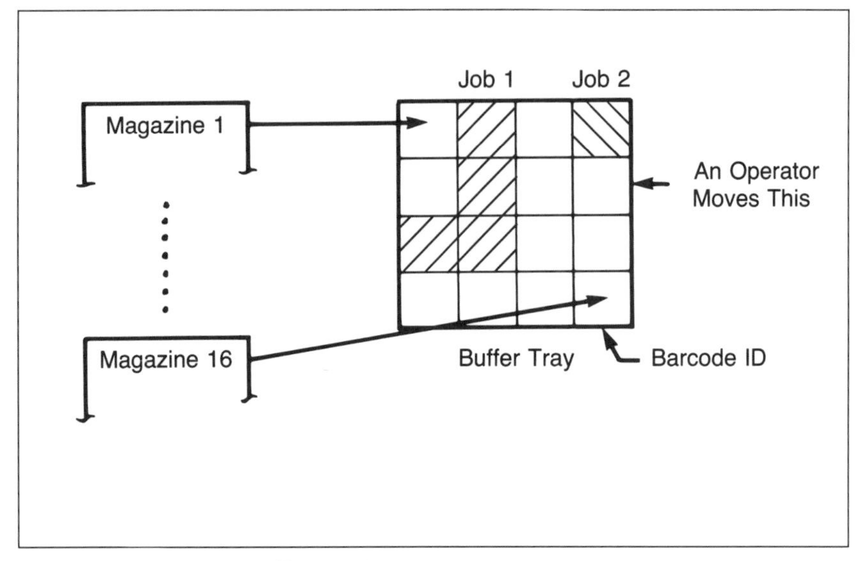

Figure 20-7. Magazine/Buffer Tray Movement.

command was performed. When providing a new magazine of parts to a tool, the Sector Controller also sends job and operational parameter update data.

After a tool has completed a magazine, it requests output magazine service from the Sector Controller and then provides the Sector Controller with completed job data and process parameters if required. The Sector Controller responds by directing the robot to move the full cassette to the tool output buffer (B) and an empty cassette to the tool output elevator.

Magazines of carriers move from tool to tool via the previously described procedure until all Robotic Line operations are completed. When the appropriate jobs have accumulated in the last tool output buffer, the Sector Controller directs an operator to remove the Buffer Tray from the last tool, replace it with a new Buffer Tray with one empty magazine, then place the completed Buffer Tray of magazines in the Output Buffer Tray Bin (OBTB). The Sector Controller provides a listing describing job related data for each Buffer Tray placed in the OBTB. The listing is in operator and barcode readable format.

Sector Controller Detailed Design

The Sector Controller was the command and control center for the Robotic Line. Hardware and software that would allow the project to eventually be maintained by manufacturing were selected. The computer used by Manufacturing at the time was the IBM Series/1.

The design of the Sector Controller allowed the tools of the sector to operate under the following conditions:

1. The Sector Controller and the Robot are operational.

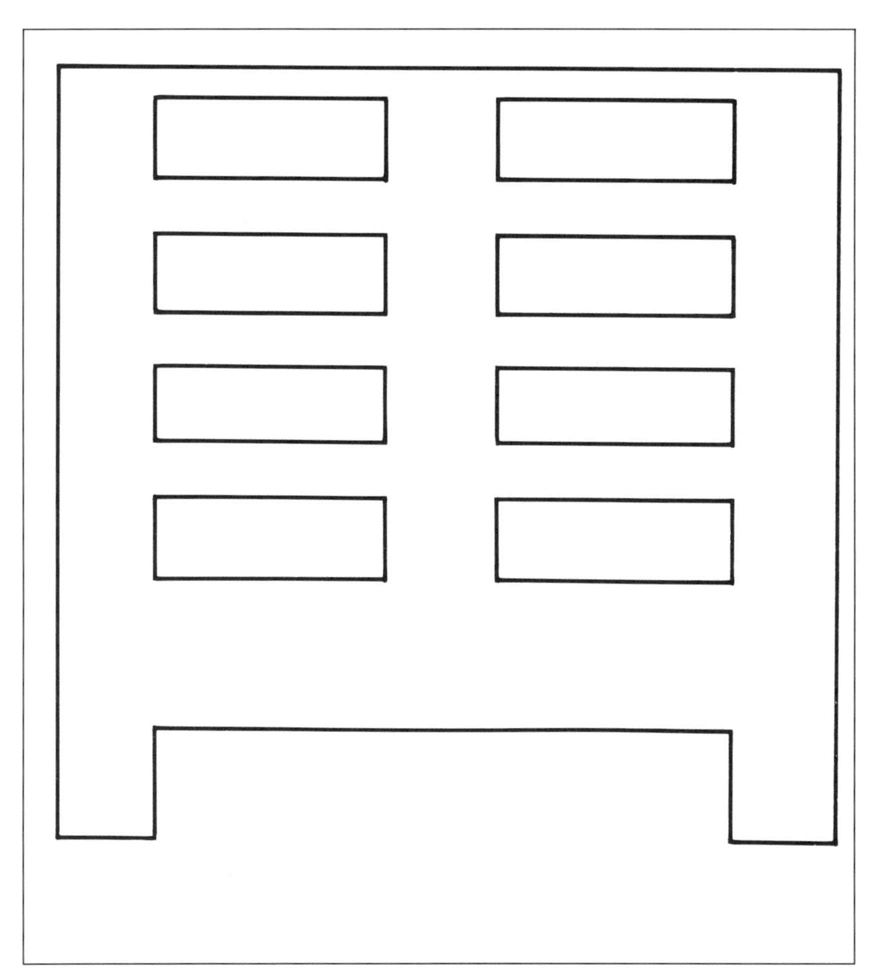

Figure 20-8. Buffer Tray Bin.

2. The Sector Controller is operational and the Robot is down.
3. The Sector Controller and the Robot are down.

Meeting this flexibility criteria required extensive programming.

Sector Controller Software Functions

The Sector Controller software operated under the Event Driven Executive (EDX) Operating system and was programmed in Event Driven Language (EDL). The major software functions are:

- Sector Configuration: This function gave qualified operators the capability to add or remove tools, buffers, or stations from the line. Unit i.d. and location data were placed in the database.

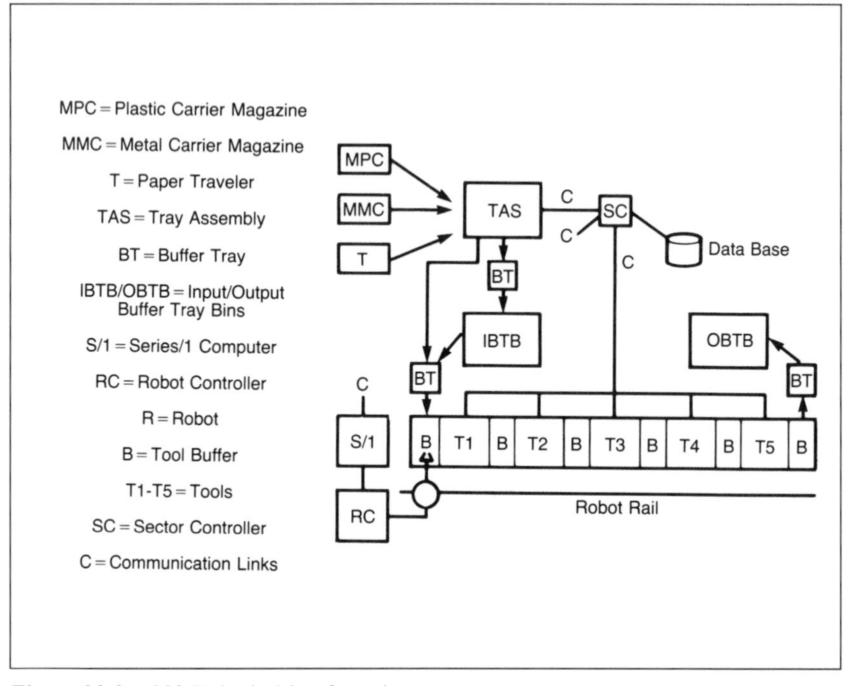

Figure 20-9. 1983 Robotic Line Overview.

- Palletizing: This function gave qualified operators the capability to enter pallet (Buffer Tray) job location data via terminal or barcode reader.
- Part Movement: This function issued directives to operators or robots to move parts to specified locations and verified that they were moved.
- Job Control: This function gave qualified operators the capability to change the priority of a job in real time.
- Parts Tracking: This function gave qualified operators the capability to locate a job, carrier, or row in real time via terminal request.
- Operational Parameter: This function gave qualified operators the capability to monitor and change tool process parameters via terminal.
- System Access: This function gave qualified operators the capability to limit system access and functions performed after access.
- Robot/Tool Utilization: This function printed out the time stamped tool service requests, robot move directives, and magazine move times.
- Sector Output Listing: This function printed out Buffer Tray content job information for each Buffer Tray Moved to the Output Buffer Tray Bin. The operations performed on each job were included in the listing.
- Sector Backup: This function printed out the data related to every magazine and tray movement and all operator and robot directives. It was used to backup the Sector Controller in the event of system failure.

Tooling Detailed Design

Five tool types were required to complete all operations of the sector. The Lab Mechanization Dept. made many high level decisions prior to starting on individual tool design. The following tool design criteria was established:

1. All tools would use common components where possible.
2. All tools would use common software where possible.
3. All tools would be designed to operate with or without the Sector Controller and Robot.
4. All tools would provide a highly visible means of displaying normal and abnormal service requests.
5. All tools would provide service personnel with effective diagnostics.
6. Tools could be purchased from an outside vendor and modified to meet line requirements if cost effective.

Typical Tool Design

Figure 20-10 shows the general layout of a typical Robotic Line tool. The Series/1 computer was used to communicate to the Sector Controller and perform high level tool logical control. High level commands were sent to the micros to perform specific I/O tasks. Sensors, monitored by the micros, were used to verify the completion of a task. The micro would send verification results to the Series/1, which in turn, would send verification data to the Sector Controller.

Most micros were common to all tools, and once coded/debugged, saved a considerable amount of time in bringing up new tools. Common micros included

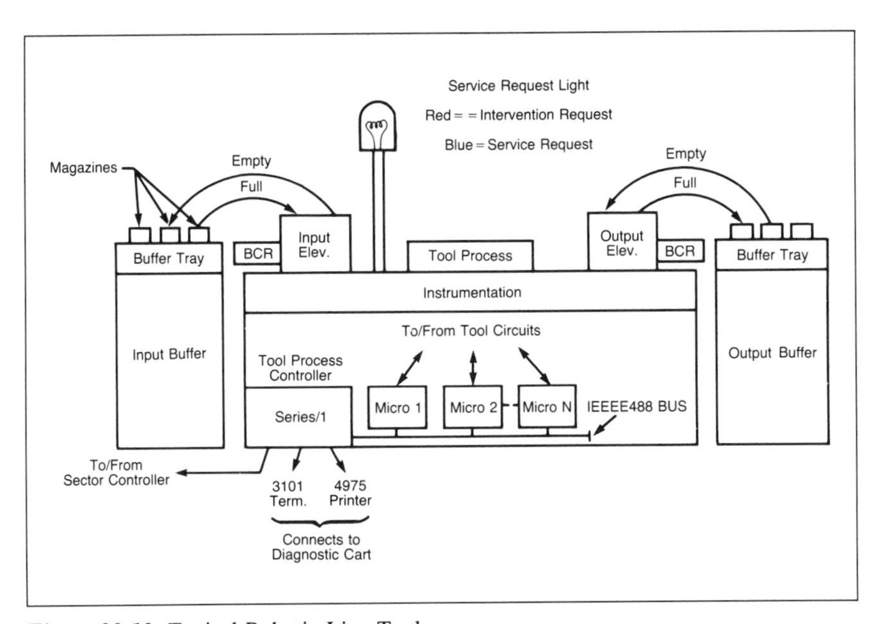

Figure 20-10. Typical Robotic Line Tool.

the elevator/barcode reader micro, the carrier move micro, the pick/place control micro, and the parameter monitor micro. Each tool also used a dedicated micro to handle tool unique I/O.

The use of distributed process control tool architecture allowed each micro to perform limited simple tasks, and to be easily coded and debugged in vendor supplied Forth.

Common hardware used by the tools included the Series/1 processor, barcode readers, elevators, micros, carrier move bar, and the service request visual indicator. All common items were available for new tool development.

Input/output buffers, that were designed to hold one removable buffer tray, were provided for each tool. This allowed all buffer trays to be interchangeable and provided the operator with a means of inputting job related magazines of parts to on-line and off-line tools.

Complex Tool Design

The most complex tool developed was the tool used to place the rows on the metal carrier. This tool performed the following tasks:

1. Input rows from a plastic seven row carrier.
2. Read the alpha numeric i.d. of the row.
3. Input a metal carrier.
4. Read the metal carrier barcode i.d..
5. Apply bonding compound to the metal carrier.
6. Place seven rows on the metal carrier.
7. Map the row/carrier i.d. data.
8. Heat the processed metal carrier.
9. Place the metal carrier in one of two cool down/align fixtures.
10. Move the empty plastic row carrier to an output elevator.
11. Remove the cooled down metal carrier/rows from the cool down/align fixture.
12. Move the completed metal carrier/rows to the output elevator.

This tool used five elevators, two barcode readers, an optical reader, several walking beam carrier move mechanisms, a small pick/place robot, a micro controlled X–Y table, and a lot of precision fixturing to mechanize the previously listed tasks. Because of the multiple operations performed by the tool, and the use of walking beam type carrier move mechanisms, as many as 20 metal carriers could be working their way through the tool at any given time.

The Series/1 programming required to operate this tool was extensive because of the multiple tasks performed by the tool and the need to send row, carrier, and magazine data to the Sector Controller. Micro commonality allowed the effort to code/debut them to be shared over several tools.

Robot Transport System

The San Jose Mechanization Dept. worked with the Special Equipment Design And Build (SEDAB) group at the IBM facility in Boca Raton to implement the robot transport system illustrated in *Figure 20-11*. This robot was programmed in A Manufacturing Language (AML) and was controlled by an

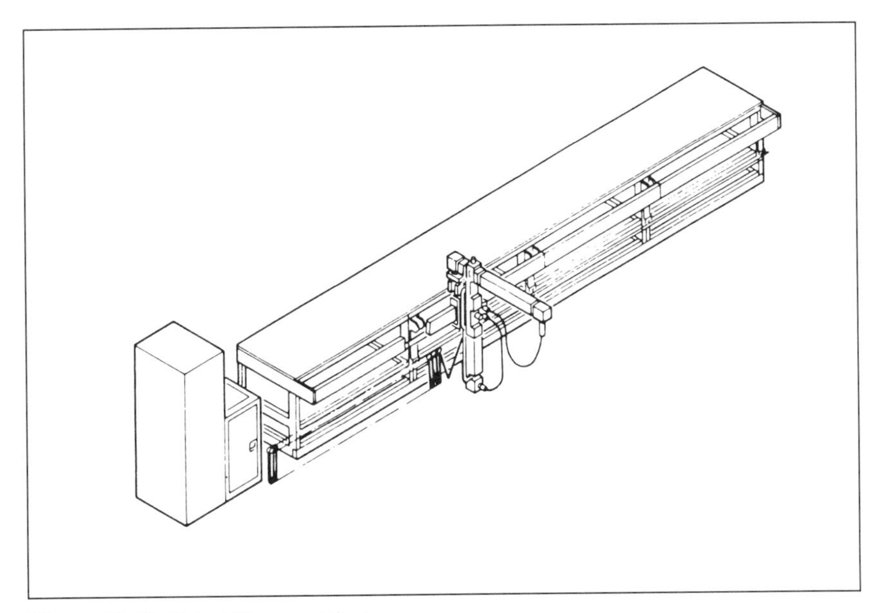

Figure 20-11. Robot Transport System.

IBM Series/1 computer. The extensibility of AML and the capability provided to write operator friendly pendant programs allowed the robot program to be written with less effort than the tooling programs.

Robot Operation

The robot moved cassettes containing a maximum of 12 carriers between tool elevators and buffers. A tool needing an input magazine would turn on the service request light then send a service request to the Sector Controller. The Sector Controller would direct the robot to remove the tools, empty input magazine, place it at a designated location in the tools input buffer tray, then pick up a full magazine from a designated location in the buffer tray and place it in the tools input elevator. After completing the directive, the robot program would send verification data to the Sector Controller.

Removal of completed magazines of metal carriers followed a similar procedure except that full magazines were removed from the tools and empty magazines were placed in the tools output elevator.

Robot System Specifications

The robot had five degrees of freedom which included X, Y, Z, Yaw and Grip. A compliance device, incorporated into the Yaw axis, provided 0.05 in. (12.70 mm) of orbital compliance. Light emitting/detecting sensors were used in the gripper to provide a means of magazine detection and buffer placement calibration.

A 16 bit absolute shaft angle encoder controlled the long Y axis and provided sufficient accuracy to move over an estimated maximum movement of 100 ft. (30.48 m). Because of the short distance to be moved and the 16 bit system control, the X, Z, W, and G axis movements were not critical, especially when compliance was incorporated into the gripper and elevator tooling. Adequate use of compliance made the robot/tool interface a non critical item.

It should not be assumed that this is a comprehensive discussion on JIT/CFM concepts. It is a brief introduction to the way these concepts can be applied to robotic line architecture. The technical books listed in the bibliography will provide the reader with a starting point for an in-depth analysis of JIT/CFM.

Background Information

After Japanese products started to impact American produced goods and after the quality of Japanese products was objectively evaluated, a lot of attention was focused on Japanese manufacturing methodology. Japanese manufacturing plants located in Japan, and subsequently in the United States, were visited and the manufacturing methodology analyzed.

In addition to the information brought back by U.S. citizens visiting Japan, some Japanese experts have also documented their manufacturing methodology and have shared it with the world.

The results of the in-depth analysis of Japanese manufacturing methodology have been documented in the form of technical books and presented to large groups of people in the form of seminars. The bibliography provided in this paper lists some of the books written by the analysts of Japanese manufacturing technology.

JIT/CFM Definition

The words "Just-In-Time" could be used to describe many conditions, but when the words "production" is added after the word "time" a clear picture is presented. The methodology of achieving this desired mode of operation is described in books like Schonberger's "Japanese Manufacturing Techniques" (1) and "World Class Manufacturing" (2) and is referred to throughout the text as JIT.

James Gooch, an IBM CFM instructor, gives the following definition for CFM. "It is an on-going examination/improvement effort which ultimately requires integration of all elements of the manufacturing process to achieve:

1. Optimally balanced line with no waste,
2. Yielding lowest possible cost on time". He also identified the CFM key ingredients as elimination of waste, production improvements, and the Kanban/Pull system.

In both JIT and CFM methodologies, a heavy reliance is put on the manufacturing discipline practiced by the Japanese. Since both have the same goals, in this paper, JIT and CFM will be used interchangeably with a slight bias toward JIT being more universally accepted.

JIT/CFM Rules For Robotic Engineers

The following rules could have been applied to the design of the 1983 Robotic Line previously described in this paper.

1. Start new manufacturing lines with minimal capital investment.
2. Eliminate operations that do not add value to the product.
3. Keep tooling as simple as possible.
4. Make tooling as reliable as possible.
5. Operate tooling at the rate that results in maximum tooling life.
6. Employ preventive maintenance for all tooling.
7. Minimize tool repair time.
8. Balance the workload of line tooling.
9. Move the product through successive tooling until completed.
10. Don't allow a defective part to move to the next tool.
11. Eliminate waste of movement of tools and product.
12. Eliminate parts buffering and off line storage.
13. Minimize data gathering, eliminate it if possible.
14. Minimize off line Quality Control.
15. Produce only what is needed, when it is needed.
16. Let the last tool in the line trigger production.

When these rules are carefully analyzed, it becomes obvious that the simple tooling may or may not include robotics or automation. While JIT/CFM proponents do not oppose the use of robots or automation, they determine whether it is suitable for the job and will not tolerate overkill.

The knowledgeable robotic engineer fully understands how he can simplify tooling by using the inherent capabilities of the robot. It is his job to show staunch JIT/CFM proponents how this can be achieved. They start from the simple and move to the complex. By definition, that means that they do not start with robots. Indeed, they may or may not have the robotic experience necessary to objectively evaluate their use.

JIT/CFM Traps

Although the application of JIT/CFM concepts have proven effective on many projects within IBM, I believe there are a few traps that may prove costly in the long run.

1. Robotics/Automation may be "planned out" of production lines because of short term planning for simple, manually operated, tooling.
2. The simple tools, initially designed for manual operation, may not be suitable for robotics or automation without major redesign.
3. Manually operated, simple tooling lines may never move to robotic or automated lines when production runs warrant it. The labor intensive, simple tooling lines will be cloned and the number of operators doubled.
4. The critical skills necessary to design/implement more complex tooling, robotics, and automation may disappear if not utilized.
5. The degree of detail required to design a product for automation may not be applied if the initial means of production is to be manually operated.

Avoiding JIT/CFM Traps

The long range thinker need not fall into these traps when planning a manufacturing line. Unless the designer knows from the start, that the production run is going to be short, initial tooling design should accommodate robotics or automation.

The following are some questions the designer should be able to answer:

1. Does the tool depend on complex vision checks? Automated/Robotized tools handle go/no-go checks better than complex vision checks.
2. Is the part being worked on easily loaded into the tool? Automated tools work best with linear, in-line feeding. Robotic tooling is more flexible.
3. How do you know the part got into the tool and left the tool? Automated/Robotized tools need go/no-go verification.
4. How do you know that the tooling operations was completed correctly? Automated/Robotized tools need simple testing built into the tooling.
5. Can the simple tools being initially designed for manual use be easily modified to mount sensors and auxiliary equipment? Automated/Robotized tools use sensors/auxiliary equipment for verification.

The list just discussed is not complete, but presents the general considerations that must be taken into account so that the designer of simple JIT/CFM tools does not design out automation/robotics.

If the manufacturing plant has a surplus headcount that must be provided with jobs, manually operated lines are acceptable as long as the cost of the labor does not make the product cost prohibitive. Producing the most reliable product at the lowest possible cost must always be the overall objective, regardless of implementation selection.

Without a doubt, JIT/CFM concepts are sound. A problem only arises when these concepts are blindly implemented in their simplest form without considerations being given to long range planning.

The last section of this paper will show the reader how the 1983 Robotic Line architecture would be improved by applying the previously discussed JIT/CFM rules. Hopefully, the designer won't fall into the traps just described.

JIT/CFM proponents are very good at asking WHY? WHY? WHY? They go at BAU head-on and don't hesitate, for one minute, to scrap BAU computerized support systems, support organizations, and tools. For this reason, conversion to JIT/CFM requires top level management approval from the start, and won't work without it. Old empires die hard.

1983 Robotic Line Design Review By A JIT/CFM Robotic Engineer

If a robotic engineer, trained in JIT/CFM methodology, had reviewed the 1983 Robotic Line Detailed Design Specification, the design would have been drastically altered. The design change recommendations would probably include the following items.

1. It's ok to automate by sector but the elaborate job control is not needed. It should be replaced by simple kanban cards.

2. The tracking of parts through the line does not add value to the product and should be eliminated. This will save the cost of barcode readers, optical character readers, and associated programming. The Sector Controller and all associated programming should be eliminated.
3. The part worked on by most tools is the metal carrier containing 7 rows. The use of 12 carrier magazines to feed tools creates excessive WIP and complicates tool design. All tools should be designed to load and process one carrier at a time. This will decrease WIP from 60 carriers (min.) holding 420 parts to 9 carriers holding 63 parts and will allow the line to operate in a Continuous Flow Manufacturing (CFM) mode. It will also save the cost of magazines, tool elevators, and associated elevator programming.
4. The tooling for the entire line should be designed for robots but should first undergo extensive manual testing. The use of robots with programmable Input/Output (I/O) capability would eliminate the need for tool processors and micros.
5. The line should produce parts only when needed. Kanban orders for parts from the last tool should trigger production, allowing the line to operate in a PULL configuration.
6. Robots used to automate tools should employ simple end effectors that do not have to be changed between operations. Changing end effectors is wasted motion and does not add value to the product.
7. Since single carriers will flow through the line and not be allowed to accumulate in magazine buffers, the Robot Transport System should not be used. Small robots used to automate the tools will move the carriers.
8. Line testing should eliminate bottlenecks, identify and correct unreliable tooling, and determine tool performance.
9. All tooling should be designed to allow go/no-go verification of tool loading, part working, and tool unloading. The small robot employed to automate the tool should use this data to control part movement. This would prevent defective parts from being passed to the next tool.
10. All tooling should be designed for ease of maintenance and ease of planned tooling changes.
11. To minimize hardware and software maintenance, all robots should be purchased from the same manufacturer and be programmed in the same language.
12. All robots should have easy to use diagnostics that pinpoint all failures to the field replaceable unit (FRU).

There are many more items that could be added to the list, but it's already apparent that the 1983 Robot Line would not meet JIT/CFM design criteria. It is also apparent that when JIT/CFM concepts are applied, savings in parts inventory, tooling, programming, and line debugging can be achieved.

The 1983 JIT/CFM Robotic Line Overview

After incorporating the changes recommended by the JIT/CFM robotic engineer, and after supportive management approved changing some old BAU practices and organizations, the Lab Mechanization Dept. may have implemented the JIT/CFM Improved Robotic Line shown in *Figure 20-12*.

The line is layed out in the shape of a large U with the robots located on the outside of the U close to wall outlets. The inside of the U would be used as the operator access when required.

Kanban control is located at the upper ends of the U; one for the input of parts at the upper left (KI), and one for the removal of finished carriers at the upper right (KO).

The advantages of this arrangements are that it minimizes the distance operators have to move, makes the robots easy to isolate from the operators, and minimizes overall floor space requirements.

A total of eight small assembly type robots (R) are used on the line. The analysis of the five original tools determined that four of the five should be broken down into two simple tools each. One tool remained unchanged and required only the load/unload functions that the adjacent tools could handle easily.

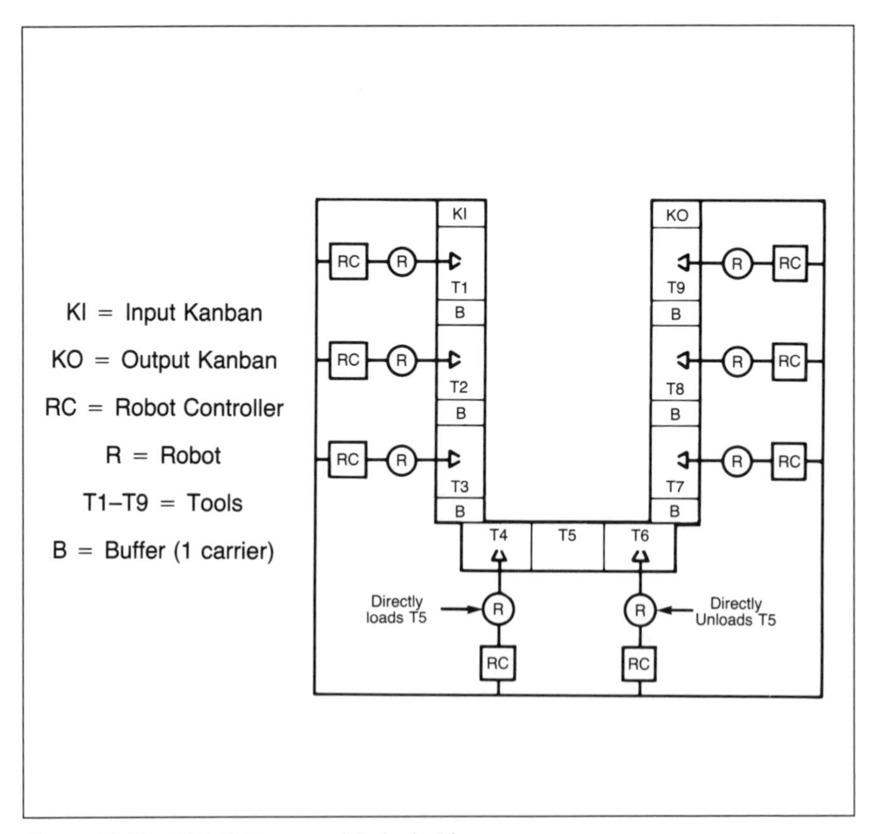

Figure 20-12. JIT/CFM Improved Robotic Line.

Parts Flow Description

This sector feeds parts to a batch process requiring a multiple number of carriers that must be loaded at the same time. When operators want parts for the batch process, they take an empty container and attached kanban card showing the required number of carriers, the location and i.d. of the tool producing the parts, and the location and i.d. of the tool using the parts to the tool producing the parts. In this case, it is the last tool of the JIT/CFM Robotic Line (T9).

The operator places the empty container in the last tool output buffer fixture (KO) and sets a count down counter, controlled by the last tool (T9), to the number of parts specified by the kanban card.

When the last tool of the line detects the non-zero count and a container in it's output buffer, it starts the production process by turning on a high visability indicator that, when illuminated, lets everyone in the area know that the robotic line should be in operation.

It then removes the single carrier from its input buffer, performs it's process on the part, places it in the empty container located in its output buffer, decrements the counter by 1, and checks for a zero count. If the count is non-zero, it repeats the process. If the count is zero, the indicator light is turned off and the tool quits producing parts.

The tool (T8) feeding the last tool (T9) detects the removal of the carrier from it's output buffer, takes the carrier from it's input buffer, performs it's operation on the part, and than places it in its output buffer.

This procedure continues, with the production of every tool being triggered by the tool that follows it. All tools would operate in this mode until the quantity specified by the output kanban was filled.

It is assumed that all tools start the process with one completed carrier in their output buffer to avoid a lag equal to the sum of the 9 tools processing time. If the tools were required to start empty, additional tool run-out logic would have to be considered.

The mechanics of how the first tool gets its parts from the preceding sector has not been discussed. This could be handled several ways.

1. When the first tool detected the carrier removed from its output buffer, it could turn on a visual indicator requesting service if it detected an empty input buffer.
2. Since the input/output kanban areas are relatively close, the operator requesting parts for the batch process could set up the first tools input buffer.

In any case, the quantity of parts placed in the first tools input buffer would be a compromise between the desired WIP and the wasted time used to transport the parts. The kanban moved between the two Sectors would reflect the quantity decided upon.

Reviewing a 1983 Robotic Line Design, and incorporating JIT/CFM concepts into a theoretical 1983 JIT/CFM Improved Robotic Line, has shown that the utilization of these concepts by robotic line designers can reduce the costs associated with parts inventory, tooling, programming, and engineering.

It has been shown that the incorporation of JIT/CFM concepts may severely impact organizations and their software support systems doing BAU. The need for top level management support has been identified.

A recommendation has been made to design all tooling, not specifically designed for short production runs, for automation and robotics. If cost effective, all tooling should be designed for automation and robotics.

The move to less complex tooling for most operations has been discussed. This requirement allows the programmable capability of robots to handle tool control previously supported by computers and associated tool programming.

The design criteria that tool controlling robots must meet has been discussed and includes reliability, serviceability, and programmability.

Finally, I recommend that all persons involved in the manufacturing process use the COMMON SENSE MANUFACTURING (CSM) concept that includes the following guidelines:

1. Be receptive to analyzing all new concepts objectively.
2. Never let Business-As-Usual (BAU) block your objective analysis.
3. Always question proponents of concepts if their way is the only way.
4. It's ok to accept some parts of a concept and reject other parts.
5. Never give up long term profits for short term strategy.
6. Expect the concepts you work with today to be replaced with better concepts tomorrow.[7]

Too frequently, the relationship between the worker and automation has been portrayed as being one of adversity, with workers being displaced as a result of automation. Richard W. Hammond, in his paper *Utilizing Manufacturing Excellence Concepts for the Successful Application of Shop Floor Technologies*, explains how the concepts of Manufacturing Excellence can be used to support every phase of the automation project.

Manufacturing Excellence (ME) concepts and methodologies include the subjects of total quality control (TQC), Just-In-Time (JIT) manufacturing and employee involvement (EI). These methodologies are generally referred to as the soft issues associated with changing the way we manufacture products. They are people/worker oriented. The three methodologies under Manufacturing Excellence are generally not conducted independently. The components of ME are generally conducted in an environment of or in combination with Employee Involvement. Considerable attention will be given in the Autofact proceedings and conference to the technical issues associated with developing the Integrated Factory. This paper will explore the human issues associated with automation projects. I will explain how ME methodologies and specifically how EI can be utilized to support the Manufacturing Engineer or Program Manager in every phase of the automation project. In other papers or documentaries concerning the worker and his relationship to automation we generally hear of the displaced operator, resulting unemployment and the final impact on society. The employee is too valuable to release or layoff for automation or the factory-of-the-future to function.

In this paper the components of Manufacturing Excellence (ME) will be defined and briefly described. A hypothetical manufacturing engineering automation project will be tracked through from inception to operation. The need for ME and specifically Employee Involvement will be tracked through the process. The paper will conclude with several case studies in an addendum, illustrating the ME methodologies.

Manufacturing Excellence

Manufacturing Excellence *(Figure 20-13)* refers to a collection of three basic concepts utilized to drive enterprises to a level of excellence. The concepts include:

- Just-In-Time Manufacturing
- Total Quality Control
- Employee Involvement

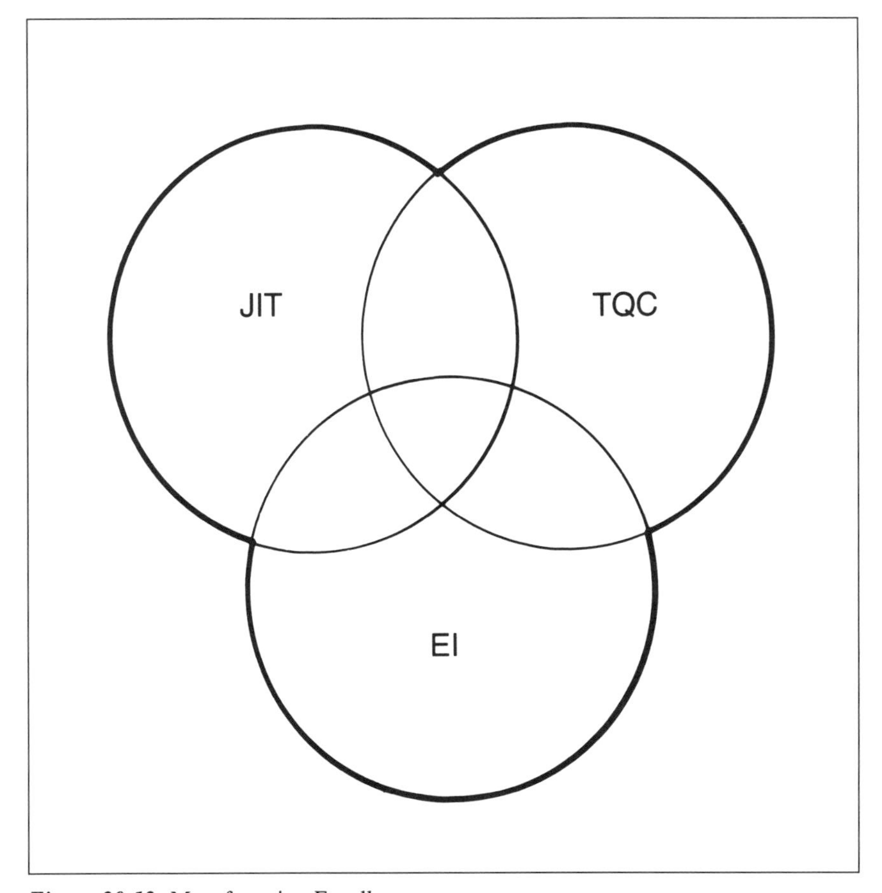

Figure 20-13. Manufacturing Excellence.

As indicated in the Introduction the concepts are generally utilized together and the pivotal concept is EI. Each of the two others, JIT and TQC will have minimal impact to an enterprise without EI. A review of the three components of Manufacturing Excellence will be provided to serve as a reference in the automation discussions.

Just-In-Time Manufacturing

Just-In-Time refers to a collection of disciplines that manufacturers adhere to in seeking or striving for excellence. The disciplines are listed in *Figure 20-14*. The basic JIT philosophy revolves around continuous improvement and the elimination of waste. The continuous improvement in an enterprise refers to the collection of incremental changes which take place to improve an operation. The incremental changes are continuous and touch on all aspects of an enterprise. In seeking continuous improvement, root causes of problems are sought and permanent solutions are put into place. The JIT concept of continuous improvement can be part of every employee's basis of doing work. Continuous improvement activities are performed from the chief executive's office to the line operator. In an environment of continuous improvement we are not relying on the "big change," "large captial project," or "the automation project" to turn the manufacturing enterprise around. The second philosphy behind JIT activity is the continuous elimination of waste. Waste is anything that does not add value to the product or service to our customer. In order to pursue this activity one needs to first identify who your customer is and then determine what that customer is seeking in the product or service. Inventory programs and as needed delivery schedules are only a small portion of the JIT philosophy. The key is continuous improvement and the elimination of waste.

Leadtime Focus
Housekeeping
Workplace Organization
Point-of-use Storage
Do your job right the first time
Short setup Times
Compact Layouts, Cellular Processes
Control By Sight
Preventative Maintenance
Flexible Organization and Workers
Supplier and Customer Networks
Employee Involvement

Figure 20-14. A list of JIT Disciplines.

Total Quality Control

The premise of Total Quality Control (TQC) fits well with the concepts of JIT. TQC promotes the adherence of quality throughout the organization, not only on the plant floor. TQC starts with the customer to analyze his needs as requirements in our product or service. These customer requirements are then disseminated throught the organization as critical success factors in satisfying the customer, increasing market share and resulting improvements in operation margin. The critical success factors focus on those administrative tasks and operational processes with the greatest impact on customer satisfaction. The tasks or processes identified become the focal point for continuous improvements. The continuous improvements take the form of process or procedure changes, capital expenditures to upgrade equipment or improved quality and value of incoming raw material or purchased stock. The methodologies of TQC start at the customer, run through all facets of the operation and extends beyond into the supplier/vendor base *(Figure 20-15)*. Customers include the final customer as well as internal customers in our enterprise. The organization as a whole services or provides products to the final customer. Everyone in the organization will generally provide a service or product for another person or department in the organization. Each individual in the organization has a customer who they must satisfy to achieve a successful operation and enterprise.

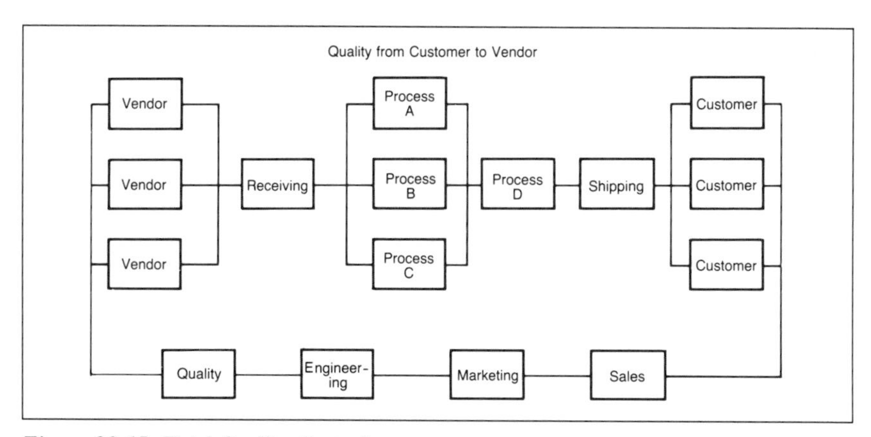

Figure 20-15. Total Quality Control.

Employee Involvement

Employee Involvement (EI) simply stated is the utilization of the creative energy of all employees in the organization to support the goals or vision of an organization. In an EI environment, ideally all employees in the organization contribute to the success of the organization through continuous improvement efforts. The EI structure is based on team activity at the functional and multidiscipline level. The teams take on the mission of identifying problems, brainstorming corrective actions, choosing a solution to the problem, then implementing a corrective action. Because the corrective action originates from the employees performing the task, the solution generally becomes a permanent part of doing business and is institutionalized. The employees at all levels, because they are involved in the change process, begin to relate closer to the goals and achievements of the organization.

Manufacturing Excellence and Manufacturing Engineering

Having reviewed Manufacturing Excellence concepts of JIT, TQC, and EI, we will discuss how a Manufacturing Engineer may utilize these concepts in the application of technology on the plant floor. For a basis of discussion the application of robots will serve as the example. Like other emerging technologies, robots received a tremendous amount of attention in the late 70s and early 80s. Engineers and managers were actively seeking out robot applications as the means to improve productivity and profitability. We will progress through the stages of an engineering project on the plant floor and specifically discuss how concepts of EI can be utilized to assure a successful program.

Stage 1—Technology Application: With emerging technologies we send engineers and sales representatives into our plants to search out the ultimate application. This technique has been utilized in applying robotics, program logic controllers, machine vision systems and voice recognition systems to name a few emerging factory floor technologies. In the case of robotics the best source of information concerning application points come from the people doing the work. The factory workers can lead us to the best applications:

- Boring jobs
- Jobs in which the worker performs his task "like a robot"
- Dangerous operations
- Uncomfortable jobs

There is no better way to gain acceptance for a new technology then to educate the workers on the new technology, then solicit ideas and potential application areas for its utilization. All the concerns about worker and union acceptance could possibly have been avoided. The application for the technology developed jointly by a knowledgeable applications engineer and effected workers has a much better potential for success.

Stage 2—Process Specifications: Once we have our potential application identified the engineer must study the operation in detail to understand the operational elements and to begin developing the Manufacturing Engineering specifications for the application of the particular technological solution. Here again the operators on the floor, performing the task are our best source of information on all aspects of an operation. No matter how simple the operation may appear during an engineering survey, only the operator is aware of the underlying trends, inconsistencies and special conditions which must be dealt with. Only from a complete engineering study of the operation, plus the input of the operators will a complete study be achieved.

Stage 3—Manufacturing Engineering Design: Now that the engineer has his application in mind and has studied the process in detail, he is now ready to plan and design the application of the automation device. Here again his design needs to be reviewed and critiqued to assure all conditions are provided for in the application. This critique needs to again include the operators on the plant floor, for a detailed review of the process to assure all contingencies are addressed. This is also a good time to involve maintenance personnel. Maintenance mechanics and electricians can express their concerns and interests in the new equipment, layout design to provide access to the equipment, and special maintenance problems associated with existing equipment. With multiple reviews, critiques and engineering changes the operators and maintenance personnel begin "taking ownership" of the new equipment representing emerging technology. The Manufacturing Engineer can now finalize his plans, layouts, and equipment choices with full confidence that all operational conditions have been reviewed. Orders can now be issued and detailed plans put in place for installation.

Stage 4—Preparation During Lead Times: At this stage of a project the employees will give the engineer additional leverage. Employee Involvement must not be abandoned. Several important project tasks need to be completed to assure a successful program. Employees need to be involved in all of the following:

- Maintenance training on new equipment.
- Equipment operational training on new equipment.
- Installation planning, in detail, to assure production is not lost for an extended period.
- Repair and modification of existing equipment to interface with the robotic device or new technology.
- Site preparation which can take place during normal production.

The items indicated above, if performed effectively will provide greater assurance of a smooth installation and subsequent operation. Employees at all levels and in multiple functions are involved in the planning and preparation of the installation, again, to become part of the project and to assume ownership.

Stage 5—Installation: This stage obviously must involve more people than the Manufacturing Engineer. By having the operations and maintenance personnel involved in the process early, their project ownership will assure a smooth installation:

- Trade and maintenance personnel will be fully familiar with machinery installation.
- Electrical interfaces will have been fully engineered and discussed.
- All aspects of the installation will be known by those individuals performing the work.

The installation will be smooth and unexpected contingencies will be minimized.

Stage 6—Startup, Operation and Maintenance: Here again, if the operators, set-up, and maintenance personnel already are involved this last stage will proceed smoothly. The Manufacturing Engineer must realize that his mission in life is to complete engineering projects and turn the installation over to the operation. The Manufacturing Engineer needs to move on to other projects and challenges. This last stage is not the place to begin training employees for the project's transition from an engineering project to operational capital. When effected employees have taken ownership of an engineering project and team up with the Manufacturing Engineer, the chances of project success without overruns is assured.

Conclusion

Manufacturing Excellence methodologies have a place on the factory floor even in the installation of the technological solution. The addendum provides several capsuled case studies to illustrate EI concepts in the application of technology in manufacturing plants. In each stage of the engineering project (See *Figure 20-16*) factory employees need to be involved to assure the sound application of technology. By keeping the employees uninformed and on the side lines, the chances of success are reduced and the Manufacturing Engineer will face a long-extended project of installation and equipment modifications, adjustments, changes, and multiple start-ups and disappointments. Only through open communication and collaboration between the trained engineer and knowledge workers will project success be assured.[8]

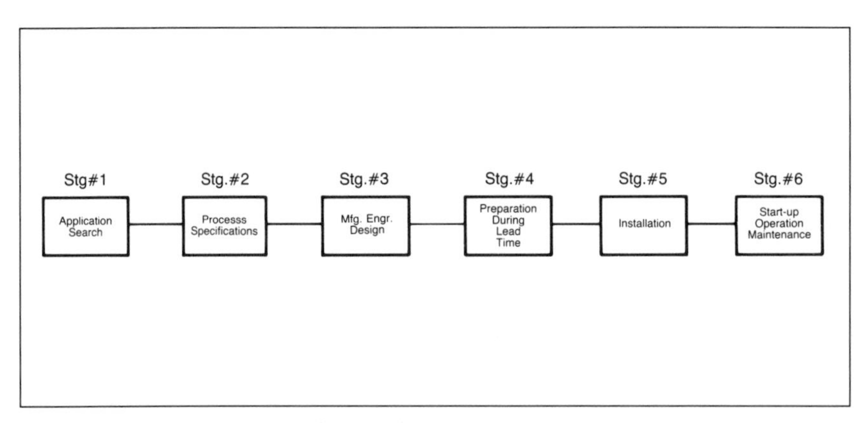

Figure 20-16. An illustration of the Project Cycle.

Mr. Hammond went on to present three case studies. They are presented in this volume as *Figure 20-17*, *Figure 20-18*, and *Figure 20-19*.

Case Study Number 1

- Fortune 100 manufacturer, operating internal engineering support effort out of the corporate research center. Engineering support effort provided manufacturing engineering services up to and including turnkey robot and machine vision installations.
- Utilization of employee involvement:
  - —Project selection involved employees.
  - —Local plant Manufacturing Engineering involvement.
  - —Frequent review of engineer plan with local plant employees and engineers.
  - —Operators and maintenance people visited research center for training in maintenance and operation.
  - —Team effort by research center and operational unit personnel.
- Results:
  - —Major increase in installation of automation equipment including robots, controls, and machine vision systems.
  - —Applications successful and still operating.
  - —Minimal deviations from expected project completion dates.

Figure 20-17. Case Study 1.

Case Study Number 2

- Major automotive supplier of drive-line components institutes JIT layouts to improve productivity and quality.
- Utilization of employee involvement:
  - —On project startup, responsible manufacturing engineer meets with "Quality of Work Life" team.
  - —Continual interchange with employees during planning stage.
  - —"Quality of Work Life" team:
    - Critiques engineer layouts and process plans.
    - Takes on part of project responsibility to rearrange tools, etc.

Results:

—JIT cells being installed with major productivity and quality improvements.

—People oriented improvements to improve the work environment also installed which improved employee morale.

—Major productivity improvements with a minimal investment in capital.

Figure 20-18. Case Study 2.

Case Study Number 3

- Automotive supplier utilizes ineffective CAD system in Tool Engineering functions. Need exists to replace the system. Preliminary engineering analysis indicates that a PC-based CAD system may be a cost-effective solution.
- Utilization of employee involvement:
  - —Tool Engineers given the opportunity to evaluate and select hardware package of PC and digitized board.
  - —Tool Engineers given the opportunity to evaluate and select software package.
- Results:
  - —Successful CAD implementation.
  - —Design as well as communication (Shop Floor, Purchasing) aspects utilized.
  - —All Tool Engineers utilize system.
  - —Database being generated for continual productivity gains.

Figure 20-19. Case Study 3.

References

1. Glenn Graham, ''Just-In-Time,'' Automation Encyclopedia: A to Z in Advanced Manufacturing, 1988, (Dearborn, MI: Society of Manufacturing Engineers).
2. Susan Lloyd McGarry, ''JIT: Helping US Manufacturers Compete,'' Manufacturing Engineering, May 1986, (Dearborn, MI: Society of Manufacturing Engineers).
3. Edward A. Herring, ''Defining Flexible Manufacturing,'' Flexible Manufacturing Systems Conference, March 1986, (Dearborn, MI: The Computer and Automated Systems Association of the Society of Manufacturing Engineers).
4. A. Thomas Jacoby, ''Manufacturing Methods Revamped at Kodak,'' May 1986, (Dearborn, MI: Society of Manufacturing Engineers).
5. David K. Doerflein, ''Just-In-Time with Multiple Gantry Robots,'' Robots 11/17th ISIR Conference Proceedings, 1987, (Dearborn, MI: Society of Manufacturing Engineers).
6. ''Marrying JIT and MRP,'' CIM Technology, Vol. 5, No. 1, (Dearborn, MI: The Computer and Automated Systems Association of the Society of Manufacturing Engineers).
7. James W. Kelley, ''The Impact of JIT and CFM Concepts on Robotic Manufacturing Line Design,'' Robots 12 and Vision '88 Conference Proceedings, 1988, (Dearborn, MI: Society of Manufacturing Engineers).
8. Richard W. Hammond, ''Utilizing Manufacturing Excellence Concepts for the Successful Application of Shop Floor Technologies,'' Autofact '88 Conference Proceedings, (Dearborn, MI: Society of Manufacturing Engineers).

A GLIMPSE OF THE FUTURE 21

As we consider likely scenarios of the future, it becomes evident that the methods of manufacturing and their end products will continue to evolve. The companies most likely to survive and prosper will be those who minimize waste in their existing operations, adequately construct and execute a waste-reducing business plan, and focus on products which capitalize on waste reduction principles.

In terms of future organization characteristics, consider these signposts of change predicted to prevail in future business by Robert Gilbreath in his book *Forward Thinking*

- More joint ventures and risk sharing
- Smaller product inventories
- More manufacturing to order
- Management emphasis on basics
- More flexible production lines
- More multidisciplinary teams
- Emphasis on internal communications
- More cross-functional orientation
- Flatter company organizations
- Management participation
- Worker participation in innovation and problem solving (quality circles)

The reader will recognize that these concepts are entirely in line with the precepts of Just-In-Time Manufacturing. Future organizations will need to be leaner, more participative, and much more responsive to customer demand. This implies flexibility, control, and maximum productivity. Specific corollaries exist for these in this text. For example:

- Increases in joint ventures and risk sharing were discussed in Chapter 7 on Advanced Procurement Technology.
- Reductions in WIP inventory were repeatedly covered, notably in Chapter 6 on Process Oriented Flow.
- Multidisciplinary teams, are covered in Chapters 11 through 18.
- Worker participation is the crux of employee involvement, as described in Chapters 10 and 18.

Without a doubt, Just-In-Time reflects a sound path for organization development and an excellent model for management structure.

In terms of product development and/or selection, JIT provides us with useful guidelines for these activities. Consider the following examples:

Microprocessors: There is a new development in the business of making microprocessors called RISC; reduced instruction-set computing. RISC involves the same kind of reasoning discussed in Chapters 5 and 6 of this book. Think of a computer as a factory which receives raw data and converts it into an end product which is useful information. The manufacturing, subassembly and assembly operations are addition, subtraction etc. which are carried out by the machines (computer chips) in our factory. Typical factory product flows, and likewise the flow of information through our computer (information factory) looks like a spaghetti bowl. Instead of presses and lathes, the computer contains a jungle of transistors, circuitry, and memory chips between which the product travels. The microprocessors themselves have become increasingly complex over successive generations of computers and chip design, in order to allow the same software to be used by any generation of microprocessor. In essence, then, we have a set of extremely complex and flexible machines organized in a poor layout, with a great deal of redundancy, excessive material (information) handling, and non-value added activity. Sound familiar?

RISC replaces the flexible equipment (complex microprocessors) with processors wired for only a few commonly used instructions, dedicated to simpler process-oriented streams. (Refer to the discussion on simple, dedicated equipment in the manufacturing environment in Chapter 5.) By utilizing this less-complicated, more dedicated processing equipment repeatedly and in varied sequences, the overall computing process is greatly simplified. The result is reduced throughput times (or improved computing speed) by up to 90%. Interestingly enough,RISC manufacturers are now struggling with the concept of synchronous production through balanced work loading. One recent magazine article referred to this problem when it stated that RISC processors can only run at full speed if surrounding circuitry can feed it data (raw material) fast enough from memory chips. This blossoming technology is expected to grow from $17 million in 1987 to a $400 million industry by 1992.[1]

A similar concept known as parallel processing is creating dramatic improvements in the same industry by bypassing bottlenecks in serial elements of problem solving. Parallel processing had already made some processing 100 times faster than previously possible by breaking up complicated problems and attacking them simultaneously. With new advances, a machine called the Ncube Model 10 has reportedly achieved speeds 1,000 times as fast. Simplification and synchronization are the keys.

One of the real time-intensive elements in the work of health care and food industry professionals is hand washing. Dirty hands have been reported to be the number one carrier of bacteria which causes diseases. We have two problems then; setup time and quality. To address these needs, a company called Pacific Biosystems has invented three systems designed to quickly, effectively clean bacteria from hands using a high-powered spray of chemicals, air, and water. Elapsed times are down to 20-90 seconds (setup time reduction), and bacteria killed is reportedly up more than 60 % over manual hand washing (quality improvement).

Anywhere there is waste, there is application for JIT (waste-elimination). Entire industries will be created and recreated in our life-times by the use of these techniques in every field. The possibilities—and the potentials—are endless.

---

[1]Herb Brody, "RISC-y Business, *High Technology Business* August 1988, vol. 8, no. 8, p.9.

CONCLUSION 22

We have learned a number of things in the course of our discussions within this text. Among the most salient points are:

The environment within which American manufacturing will be competing poses new threats, and new opportunities. The kind of organization, manufacturing technology, and business philosophy required to meet the challenges and exploit the opportunities of the future is one of continuous improvement and waste elimination. Cornerstones of the new manufacturing structure will include enhanced quality levels in both products and services, higher levels of employee involvement in day-to-day operations decisions, a commitment to continuous improvement at every level, an interdisciplinary teaming approach to management, and an ability to constantly evaluate and incorporate changes into the business itself.

A number of approaches have been offered to achieve all or portions of these goals, from requirements planning systems through artificial intelligence. Many of them are valid approaches to subsets of the required improvements. However, a number of limitations and disadvantages apply to most of these approaches; some are planning-only, some are cost-prohibitive, and still others involve technology which has yet to be developed fully. Just-In-Time offers an approach to manufacturing excellence which is:

Timely—As we have seen in the earliest chapters of the text, a majority of U.S. manufacturers have yet to embrace JIT philosophies in a meaningful way, and have need of the tremendous advantages offered by them.

Proven—Through our case studies, and in instances cited earlier, JIT has been repeatedly proven in American manufacturing settings from high-tech to low-tech, job shop to multi-national conglomerate.

Powerful—JIT implementations have been widely documented (and referenced here) to reduce operating costs, inventory levels, throughput times, and defect levels dramatically.

Cost-effective—Especially when it is compared to approaches such as flexible manufacturing systems, JIT is a low-cost means to achieve critical success factors.

Major components of the JIT philosophy, as applied in the manufacturing environment, include quality, simplified synchronous production, process oriented flow, advanced procurement technology, improved design methods, enhanced support functions, and employee involvement.

The major steps involved in a sound JIT implementation include program organization, opportunity assessment, pilot planning and design, pilot implementation, remaining facility planning and design, remaining facility implementation, program monitoring/continuous improvement, and an ongoing employee involvement program.

We have covered the content of each of these concepts and processes, and discussed their interrelationships. We have reviewed case study data from both current publications and the author's own experience. We have seen models of many of the principles involved, as well as for the program itself.

Finally, we have reflected briefly on the application of JIT thinking in diverse and future oriented products. We have looked at the way organizations must evolve to facilitate innovation and manage change in their own environs, so that simplification techniques can be merged with technological advances to provide the best of both worlds. As American manufacturers, from design through distribution, this must be our inevitable goal.

INDEX

D

E

F

O

P

Q

V

W

Y

Z